D0931679

460
TEN

Lecture Notes in
Operations Research and
Mathematical Systems

Economics, Computer Science, Information and Control

Edited by M. Beckmann, Providence and H. P. Künzi, Zürich

33

K. Hinderer
Institut für Mathematische Stochastik
der Universität Hamburg

DUNB

Foundations of Non-stationary Dynamic Programming with Discrete Time Parameter

Springer-Verlag
Berlin · Heidelberg · New York 1970

Advisory Board
H. Albach · A. V. Balakrishnan · F. Ferschl
W. Krelle · N. Wirth

This work is subject to copyright. All rights are reserved, whether the whole or part of the material is concerned, specifically those of translation, reprinting, re-use of illustrations, broadcasting, reproduction by photocopying machine or similar means, and storage in data banks.

Under § 54 of the German Copyright Law where copies are made for other than private use, a fee is payable to the publisher, the amount of the fee to be determined by agreement with the publisher.

© by Springer-Verlag Berlin · Heidelberg 1970. Library of Congress Catalog Card Number 72-132274 Printed in Germany.
Title No. 3782

T
57.83
H55

Preface

The present work is an extended version of a manuscript of a course which the author taught at the University of Hamburg during summer 1969.

The main purpose has been to give a rigorous foundation of stochastic dynamic programming in a manner which makes the theory easily applicable to many different practical problems. We mention the following features which should serve our purpose.

a) The theory is built up for non-stationary models, thus making it possible to treat e.g. dynamic programming under risk, dynamic programming under uncertainty, Markovian models, stationary models, and models with finite horizon from a unified point of view.

b) We use that notion of optimality ($\bar{p}$-optimality) which seems to be most appropriate for practical purposes.

c) Since we restrict ourselves to the foundations, we did not include practical problems and ways to their numerical solution, but we give (cf.section 8) a number of problems which show the diversity of structures accessible to non-stationary dynamic programming.

The main sources were the papers of Blackwell (65), Strauch (66) and Maitra (68) on stationary models with general state and action spaces and the papers of Dynkin (65), Hinderer (67) and Sirjaev (67) on non-stationary models. A number of results should be new, whereas most theorems constitute extensions (usually from stationary models to non-stationary models) or analogues to known results.

Our treatise is far from being complete. In particular, we only treated the expected cost criterion, thus regrettably excluding the interesting work of Derman, Veinott and others that has been done on the average cost criterion and on small interest rates in stationary models.
We also did not treat the relations between dynamic programming and the most interesting and deep work of Dubins and Savage (65).

The author is indebted to Dr.M.Schäl for reading the manuscript and making a number of valuable suggestions, to Dr.G.Hübner for a careful revision of the manuscript and for drawing the figures, and to Mrs.E.Schmidt for her diligent typing.

Hamburg, May 1970 K.Hinderer

Contents

1. Introduction and summary

The literature on stochastic dynamic programming has grown in the last 10 years rather vigorously. After the pioneering work of Bellman, Howard and others, the papers of Blackwell (65) and of Dynkin (65) were most important for the further development of the foundations, as far as the expected cost criterion is used. Blackwell was the first who gave a rigorous treatment for non-countable state spaces, whereas Dynkin studied non-stationary models (with countable state space). Another important paper for non-stationary models is Sirjaev (64). Strauch (66) extended the results of Blackwell to other stationary models. Hinderer (67) generalized the results of Blackwell (65) and Dynkin (65) to non-stationary models with general state space. Sirjaev (67) is a very useful review paper, which treats also the case of continuous time parameter. The most general sufficient condition for the existence of optimal policies (or plans, as we shall say) is due to Maitra (68).

Our treatise is divided in two chapters. The first one treats models with countable state space (and arbitrary action space), whereas in the second one we investigate models where the state space and action space are Borel subsets of complete separable metric spaces. All features of the optimization problem are already developped in chapter I, thus making it possible to concentrate in chapter II on the measure-theoretic aspects of the problem. We feel that this division is appropriate from the didactical point of view. On the one hand, readers with little knowledge in measure theory will be able to understand all proofs of chapter I and by reading essentially only the definitions, theorems and remarks of chapter II, they should be able to apply the results of that chapter too. On the other hand, readers with a sound background in measure theory will probably find it useful to have in chapter I an introduction to the essentials of the problem before studying it in its full generality.

Now we are going to summarize the contents of the different sections and to give some comments on it.

We start by defining in section 2 the essentials of a problem of non-stationary dynamic programming under risk: the (countable) state space S; the space of actions A; the set $D_n(h)$ of actions available to us if at time n the history $h=(s_1,a_1,\ldots,s_n)$ has occurred; the initial distribution $p_o(s_1)$ and the probabilistic transition law $p_n(h,a,s)$ which is the probability to be in state s at time n+1, if at time n the history h has occurred and if we take action a; the reward $r_n(h,a)$ given to us at time n+1. Any problem may be defined in terms of an appropriate tupel $(A,S,(D_n),(p_n),(r_n))$. There follow the definitions of a plan and the construction of the decision process generated by a plan. In order to ensure the convergence of the total reward and the existence of the expected reward under any plan f, we make two assumptions which seem to be rather general and most natural (as long as one does not care about the specific transition law): Either all reward functions are "essentially" positive (case (EP)) or all reward functions are "essentially" negative (case (EN)). As in the work of Blackwell and Strauch, it turns out that sometimes the two cases lead to quite different results. Particularly clear is the situation when case (EP) as well as case (EN) holds. This so-called case (C) is the natural generalization of the discounted stationary case. At the end of section 2 we discuss the notion of $\bar{p}$-optimality, used throughout the present work instead of the usual notion of optimality.

In section 3 we make some effort to elucidate the relation between the famous principle of optimality and the optimality equation (OE). It might be surprising that according to our study the dominant role, usually ascribed to the principle (as it stands, cf. theorem 3.8) seems rather to be due to the OE which serves to derive the important optimality criterion (theorem 5.1). The representation of the expected reward by means of the operators Λ_{nf} (cf.lemma 3.6) proves very useful in several instances, as well as the representation of the maximal expected reward by means of the operators U_n (cf.theorem 4.1).

Section 5 contains a number of optimality criteria, the most useful of which (theorem 5.11) is a generalization of a criterion of Maitra (68). This criterion goes much beyond the usual assumption that the sets $D_n(h)$ of admissible actions are finite. In section 6 we develop the important concept of a sufficient statistic and derive from earlier results several important theorems for models with sufficient statistics. Afterwards we specialize to Markovian and stationary models. Section 7 is devoted to so-called models with incomplete information, widely used in electrical engineering. These are models where: (i) the history consists of an observable part h and a concealed part z, (ii) $D_n(h,z)$ does not depend on z, and (iii) only those plans are used which do not take into account the concealed history. It turns out that such models with incomplete information may be reduced to models in our sense. In section 8 we represent a number of examples. We emphasize the structural properties of these examples, since many interesting practical applications are easily available in the literature (e.g. Aris (64), Beckman (68), Bellman (57), Bellman and Dreyfus (62), Boudarel, Delmas et Guichet (68), Jacobs (67), Künzi,Nievergelt und Müller (68), Nemhauser (66), Neumann (69), White (69)).

Randomized plans are considered in section 9. It turns out that we can dispense with randomized plans in case (EN), but not in case (EP). Chapter I closes with an analysis of dynamic programming under uncertainty by means of the Bayesian approach. Our main object is to show how to reduce that problem to a problem of dynamic programming under risk. Having this result established, we may easily derive many results from previous sections by means of a suitable sufficient statistic.

Chapter II is mathematically much more sophisticated than chapter I. It is clear from Blackwell's work that a satisfactory theory cannot be developped for completely arbitrary state space and action space. Fortunately, in applications, the state space and the action space are usually either countable sets or subsets of euclidean n-space. We decided to work within the frame of the theory of Blackwell, hence admitting that the state space and action space are Borel subsets of complete separable metric spaces (so-called SB-spaces). One might ask why we did not restrict ourselves to subsets of euclidean n-space. The main reason is that such a restriction would not result in any substantial simplification of the proofs, whereas our assumptions include the case of euclidean n-space as well as the countable case and the case of some function spaces. Readers who are interested to use the results of chapter II as a rigorous basis for practical problems, but who have only a rudimentary background in measure theory and topology, are advised to read only the definitions, theorems and remarks of sections 11, 14, 15, 17 and particularly 18, and to replace "SB-space" by "subset of Euclidean n-space", "universally measurable function" and "measurable function" just by "function". On the other hand, the notion of an upper semi-continuous real function must be taken in its rigorous meaning. - In order not to tire the reader, we did in general not repeat arguments already used in chapter I.

Section 11 parallels section 2 in so far as therein the general foundation for decision models is given. In section 12 we collect the measure-theoretical and topological material needed in later sections. Section 13 contains the generalization of the result of Strauch on the universal measurability of the maximal conditional expected reward. In the following sections we carry through an analysis of the optimization problem, very similar to that made in chapter I. The sections 14, 15, 17 and 18 correspond to sections 3+4, 9, 5 and 6, respectively. In section 16 we generalize the fixed point theorem for contractions to sequences of operators, which proves to be useful in sections 17 and 19. We conclude by supplementary remarks concerning other notions of optimality.

The references in the statements of the theorems should not be taken as statements of priority but rather as an indication of the sources we have used. Of course, we tried to refer to the original sources, but we do not think that we succeeded herein completely since many basic ideas of dynamic programming are scattered over a large number of papers.

Chapter I. Countable state space

2. Decision models and definition of the problem.

At first we give a formal definition of the model and afterwards a series of remarks.

The mathematical framework for a problem in dynamic programming under risk with a countable state space consists of a tupel $(S,A,D,(p_n,n\in\mathbb{N}_o),(r_n))$ of objects of the following meaning and intuitive interpretation.

(i) S is a non-empty countable set, the so-called *state space*.

(ii) A is a non-empty set, the so-called *space of actions*. We shall use the set $\bar{H}_n := S\times A\times S\times\ldots\times S$ (2n-1 factors) of *histories* $h_n=(s_1,a_1,s_2,\ldots,a_{n-1},s_n)$.[+)]

(iii) D is a sequence of maps D_n from certain sets $H_n\subset\bar{H}_n$ to the set of all non-empty subsets of A with the property that

$$H_1 = S ,$$

$$H_{n+1} = \{(h,a,s):\ h\in H_n, a\in D_n(h), s\in S\},\quad n\in\mathbb{N}. \tag{2.1}$$

$D_n(h)$ is called the *set of admissible actions* at time n under history h, whereas H_n is called the set of *admissible histories* at time n. We shall denote by K_n the set $\{(h,a): h\in H_n, a\in D_n(h)\}$, hence $H_{n+1}=K_n\times S$. The sets H_n may conveniently be regarded as parts of the infinite tree H which consists of those sequences $(s_1,a_1,s_2,\ldots)$ for which $(s_1,a_1,\ldots,s_m)\in H_m$ $\forall m\in\mathbb{N}$. (cf.fig.1).

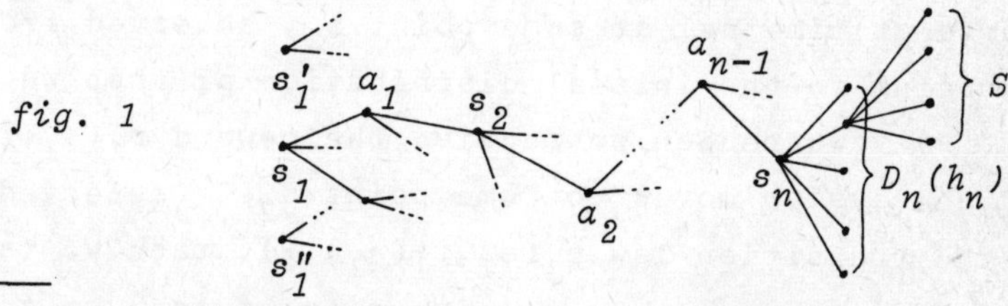

+) In order to simplify the rather involved notation we shall in the sequel suppress the time index ν in s_ν, a_ν, and h_ν whenever it is clear from the context which number ν is meant. Furthermore, s_ν and a_ν without further specification will denote the $(2\nu-1)$th and (2ν)th element of the sequence h_n under consideration. Finally, for any $h_n\in\bar{H}_n$ we shall denote by y_ν the sequence $(s_1,s_2,\ldots,s_\nu)$ of states up to time ν contained in h_n.

(iv) p_o is a counting density+) on S, the so-called *initial distribution*++). We shall usually write p instead of p_o. $p_n(h,a,\cdot)$ is a counting density on $S, n \in \mathbb{N}, (h,a) \in K_n$. $p_n(h,a,\cdot)$ is called the *transition law* between time n and n+1.

(v) r_n is an extended real valued function on K_n, the so-called *reward* during the time interval $(n,n+1>$.

Definition. Any tupel $(S,A,D,(p_n),(r_n))$ with the properties (i)-(v) stated above is called a (stochastic dynamic) *decision model* (abbreviated by DM).

A DM comprises all data necessary to define a dynamic optimization problem. Yet there remains the task to define the notion of a plan (otherwise often called a policy) and (under suitable assumptions on (r_n)) of a criterion of optimality.

Definition. A (deterministic admissible) *plan* is a sequence $f=(f_n)$ of maps $f_n : S^n \to A$ with the property

(2.2) $$f_n(y) \in D_n(h_{nf}(y)), n \in \mathbb{N}, y \in S^n,$$

where

(2.3) $$h_{nf}(y) := (s_1, f_1(s_1), s_2, f_2(s_1,s_2), \ldots, f_{n-1}(s_1,s_2,\ldots,s_{n-1}), s_n)$$

denotes the history at time n obtained by the use of plan f when the sequence $y=(s_1,s_2,\ldots,s_n)$ of states occurred.

The (non-empty) set of all admissible plans will be denoted by Δ. The application of any plan $f \in \Delta$ generates a stochastic process, the *decision process* determined by f. This process may be described verbally as follows. A system starts at time n=1 at some point s_1, selected from S according to the initial distribution p; then we take action $f_1(s_1)$ whereupon we receive the reward $r_1(s_1,f_1(s_1))$ and the system moves to some point $s_2 \in S$, selected according to the transition law $p_1(s_1,f_1(s_1),\cdot)$; then we take action $f_2(s_1,s_2)$ whereupon we receive $r_2(s_1,f_1(s_1),s_2,f_2(s_1,s_2))$ and the system moves to some point s_3 selected according to $p_2(s_1,f_1(s_1),s_2,f_2(s_1,s_2),\cdot)$, etc.

+) For the terminology, cf. appendix 1.

++) Strictly spoken, p_o is not a distribution, but it defines a distribution.

The mathematical model for the decision process determined by f will be a *probability space* $(\Omega, \mathcal{F}, P_f)$ and a sequence (ζ_n) of random variables on it, where ζ_n describes the state of the system at time n.

1) As sample space Ω we take $S^{\mathbb{N}}$, the set of all sequences $\omega=(s_1, s_2, \ldots)$.
2) As σ-algebra $\mathcal{F}$ on Ω we take the infinite product-σ-algebra determined by the factors $\mathcal{P}(S)$=system of all subsets of S.
3) ζ_n will be the n-th coordinate variable, i.e.

(2.4) $$\zeta_n(s_1, s_2, \ldots) := s_n, n \in \mathbb{N}.$$

4) For the description of P_f we shall use the notation

(2.5) $$p_{nf}(y,s) := p_n(h_{nf}(y), f_n(y), s), n \in \mathbb{N}, (y,s) \in S^{n+1};$$

i.e. p_{nf} is the transition law between time n and n+1 that results from the application of f. (Note that $(h_{nf}(y), f_n(y)) \in K_n$, $n \in \mathbb{N}$, whenever $f \in \Delta$ and $y \in S^n$.) Furthermore η_n will denote the random 'vector' $(\zeta_1, \zeta_2, \ldots, \zeta_n)$, describing the state history at time n.

It is well-known (theorem of Kolmogoroff or theorem of C.Ionescu Tulcea, cf. Loève (60),p.137) that there exists a unique *probability measure* P_f on $\mathcal{F}$ that satisfies

(2.6) $$P_f(\eta_n = y) = p(s_1) p_{1f}(s_1, s_2) \ldots p_{n-1,f}(s_1, s_2, \ldots, s_n),$$
$$n \in \mathbb{N}, y = (s_1, s_2, \ldots, s_n) \in S^n.$$

P_f may also be described as the unique probability measure on $\mathcal{F}$ such that $P_f(\zeta_{n+1} = s | \eta_n = y)$ equals $p_{nf}(y,s)$ whenever $P_f(\eta_n = y) > 0$, and such that $P_f(\zeta_1 = s) = p(s), s \in S$.

Obviously a DM may be regarded as a family of decision processes, the family being indexed by the plans $f \in \Delta$.

It remains to define a criterion of optimality. We shall restrict ourselves to the expectation of the total reward, though other criteria have been dealt with in the literature (cf.section 20). We have to make assumptions on the sequence of rewards that guarantee the existence of the expected total reward under any plan f. Most natural seem to be the following two assumptions (EN) and (EP), in the formulation of which we use $\|u\| := \sup_{x \in X} |u(x)|$ for any extended real valued function u defined on any set X.

Assumption (EN) which we call the *essential negative case:*

(2.7) $$\sum\|r_n^+\| < \infty.$$

Assumption (EP) which we call the *essential positive case:*

(2.8) $$\sum\|r_n^-\| < \infty.$$

It is easily seen that the cases (EN) and (EP) hold simultaneously iff

(2.9) $$\sum\|r_n\| < \infty.$$

We shall then speak of the *convergence case (C)*. Assumption (EN) and (EP) holds e.g., if all functions r_n are negative or positive, respectively. Case (C) holds e.g. if there are constants $c>0$ and $\beta\in(0,1)$ such that r_n is bounded by $\beta^{n-1}c$. In case (EN) and (EP) any function r_n is bounded from above or from below, respectively.

From now on we shall use the

General assumption: Either case (EN) or case (EP) holds. Our general assumption implies the existence of the expected total reward G_f under plan f for *any* transition law (p_n). Theorem A3 opens the possibility to weaken the general assumption by taking into account the specific transition law. We shall not elaborate this idea in the present work.

If we use plan f and if the sequence $y\in S^n$ of states has occurred then we receive during the time period $(n,n+1\rangle$ the reward

$$r_{nf}(y) := r_n(h_{nf}(y), f_n(y)).$$

If the sequence $\omega=(s_1,s_2,\dots)\in\Omega$ occurs then we receive the *total reward*

$$R_f(\omega) = \sum r_{nf}(y_n) = \sum r_{nf}\circ\eta_n(\omega).$$

R_f is well-defined as an extended real valued random variable on $(\Omega,\mathcal{F},P_f)$ whenever case (EN) or (EP) holds (cf.appendix 2B). From (2.7)-(2.9) and lemma A1 we get

(2.10) $$R_f^+ \le \sum r_{nf}^+ \le \sum\|r_n^+\| < \infty \text{ in case (EN)},$$

(2.11) $$R_f^- \le \sum r_{nf}^- \le \sum\|r_n^-\| < \infty \text{ in case (EP)},$$

(2.12) $$|R_f| \le \sum\|r_n\| < \infty \text{ in case (C)}.$$

Hence there exists the *expected total reward*

(2.13) $$G_f := E_f R_f = \int R_f \, dP_f,$$

and

$$(2.14) \qquad -\infty \leq G_f \leq \sum\|r_n^+\| < \infty \text{ in case (EN)},$$

$$(2.15) \qquad -\infty < -\sum\|r_n^-\| \leq G_f \leq \infty \text{ in case (EP)},$$

$$(2.16) \qquad |G_f| \leq \sum\|r_n\| < \infty \text{ in case (C)}.$$

Definition. The plan $f^* \in \Delta$ is called $\bar{p}$-*optimal* (i.e. optimal in the mean with respect to p) if

$$G_{f^*} = \sup_{f \in \Delta} G_f =: G .$$

G is called the *maximal expected total reward.*

Obviously the relations (2.14)-(2.16) remain valid if G_f is replaced by G. Of course, in cases (EN) and (EP) only those models are of interest for which $G_f > -\infty$ for at least one $f \in \Delta$, or $G_f < \infty$ for all $f \in \Delta$, respectively. We do not formally exclude these cases since this would not lead to any simplification of the theory.

Remarks. 1) We begin with some *important* remarks on the concept of $\bar{p}$-optimality of a plan.

In most practical problems there is a fixed state $x \in S$ in which the system starts with probability one. Let us call for the moment in such a model (where $p(x)=1$) a plan *x-optimal* if it is $\bar{p}$-optimal. It is characteristic of dynamic programming problems that the optimization of G_f leads after the first step to a whole family of similar optimization problems whose initial states are those states to which the system may move with positive probability under the action $f_1(x)$ taken in the first step. This fact seems to be the reason why it became customary to call a plan *optimal* if it is x-optimal for any initial state $x \in S$ (for countable S, irrespective of the actual initial distribution, cf.e.g. Howard (60); for arbitrary S cf. Blackwell (65)). This definition is on the whole (though not always, cf.(ii) below) appropriate for models with countable S since one can "compose" an optimal plan if there exists for any $x \in S$ an x-optimal plan. The situation is quite different if S is not countable because then the "composition" of x-optimal plans may produce decision rules f_n that are not measurable and hence not usable as elements of a plan. Hence, as D. Bierlein, Karlsruhe, pointed out in a discussion, the usual concept of optimality contains a uniformity condition (with respect to the initial state) that is too strong for practical applications unless S is countable. This is the reason why we decided to use throughout in the present work the notion of $\bar{p}$-optimality.

Let us mention some of its advantages when dealing with practical problems.

(i) It follows from theorem 3.8: If S is countable and if $p(s)>0$ for all $s\in S$, then the notions of $\bar{p}$-optimality and optimality coincide. Hence, if S is countable, results for optimal plans are special cases of results on $\bar{p}$-optimal plans.

(ii) Of course, any optimal plan is also $\bar{p}$-optimal. There are cases however (even if S is countable) in which no optimal plan exists but in which a $\bar{p}$-optimal plan exists. This results from the fact that optimal plans usually take into consideration states that do never occur with positive probability. Thus the notion of $\bar{p}$-optimality avoids redundant optimization.

(iii) Assume $G<\infty$. If no $\bar{p}$-optimal plan exists, then it is most important from the practical point of view that there obviously exists for any $\varepsilon>0$ a $\bar{p}$-ε-optimal plan, i.e. a plan f such that $G_f\geq G-\varepsilon$. On the other hand, Blackwell (65) has shown that if S is uncountable, then there need not exist for any $\varepsilon>0$ an ε-optimal plan, i.e. a plan f such that $G_{1f}(s)\geq G_1(s)-\varepsilon, s\in S$, where the letter s refers to the model with $p(s)=1$ (cf.section 3).

(iv) Let the model be Markovian (cf.section 6) and assume that we have found some 'good' plan π. Then one can find a Markov plan σ which is at least as good as π, i.e. for which $G_\sigma\geq G_\pi$ (cf.theorem 18.1). In particular, if there exists a $\bar{p}$-optimal plan, then there exists a Markov $\bar{p}$-optimal plan. On the other hand, Blackwell (65) gave an example where a non-Markovian plan π exists such that no Markov plan σ satisfies $G_{1\sigma}(s)\geq G_{1\pi}(s), s\in S$; and it seems to be unknown if the existence of an optimal plan always implies the existence of an Markov optimal plan.

We should like to emphasize that the foregoing remarks are solely concerned with the usefulness of different notions of optimality for *practical* purposes. They do not detract anything of the very interesting *mathematical* studies that have been done on other notions of optimality and which sometimes sharpen results on $\overline{p}$-optimality. Moreover, it will become apparent in chapter 2 that many results on $\overline{p}$-optimality may be derived by methods that were originally developped (by Blackwell, Strauch, Maitra and others) for other notions of optimality.

The notion of $\overline{p}$-optimality seems to occur for the first time (implicitely) in Strauch (66), theorems 4.1 and 4.3. It occurs in Hinderer (67) under the name of M-p-optimality. In case $-\infty<G<\infty$ it coincides with the notion of p-optimality, introduced by Blackwell (65) (cf.section 20).

2) In applications usually the set A either is countable or some (Borel) subset of $\mathbb{R}^n$. But one can imagine of applications in control theory where the actions are e.g. curves in $\mathbb{R}^3$, so that A is a certain function space.

3) There may be applications (cf.White (69),p.23) where at the times n=1,2,... different state spaces and /or different action spaces are needed. This would not complicate the theory (cp.Dynkin (65) and Hinderer (67)) but the notation becomes a bit cumbersome. Moreover, the case of time dependent state spaces and action spaces may formally - though for practical purposes somewhat artificially - be reduced to our model by using the direct sum (cf.Bauer (65), p.117) of the state spaces and the direct sum of the action spaces, and extending the definition of $D,(p_n),(r_n)$ in an obvious manner.

4) One should note that case (EP) cannot be reduced to case (EN) by 'changing signs' since in *both* cases we want to maximize G_f. Of course, problems of minimization are immediately reduced to our model by 'changing signs'. The two cases (EN) and (EP) may be divided in three disjoint cases: (i) case (C), (ii) $\sum \|r_n^+\|<\infty,\sum\|r_n^-\|=\infty$, (iii) $\sum\|r_n^+\|=\infty,\sum\|r_n^-\|<\infty$. Case (ii) usually occurs in a context where one has some

loss function $r'_n \geq 0$ and one wants to minimize the expected total loss $\sum r'_n$.

5) The cases (EN) and (EP) seem to cover all other assumptions made in the literature with two exceptions:
(i) Sirjaev (67) does not require the convergence of the series $\sum r_{nf}$ of rewards but the convergence of the series $\sum E_f r_{nf} \circ \eta_n$ of expected rewards (which in our model is a consequence of either of the assumptions (EN) and (EP), cf. theorem A4 in appendix 2). In our opinion the possible non-existence of the total reward $R_f = \sum r_{nf} \circ \eta_n$ renders the empirical interpretation and hence the applicability of the model somewhat questionable.+)
(ii) Hinderer (67) derives some results under the assumption that $\sum r_n \circ \eta_n$ converges on H and that $\sup_{m,n \in \mathbb{N}} \| \sum_{\nu=m}^{m+n} r_\nu \| < \infty$. But most results in that paper are proved under stronger assumptions (mostly case (C)) which are fulfilled in cases (EN) and (EP).

6) Sometimes the reward between times n and n+1 depends also on s_{n+1}: $r'_n(h,a,s)$. One can reduce this case easily to our model by using $r_n(h,a) := \sum_s r'_n(h,a,s) p_n(h,a,s)$ as new reward function. Another device for the reduction to our model would consist in using $r_1 :\equiv 0$ and $r_n(h,a) := r'_{n-1}(h)$. But this procedure has the disadvantage that Markovian models do not necessarily remain Markovian. On the other hand, there do not arise complications in building up a theory with r'_n depending also on s_{n+1} (cf.e.g.Blackwell (65), Strauch (66), Hinderer (67)). We have choosen r'_n independent on s_{n+1} for reasons of simpler notation.

7) In some papers (e.g. Sirjaev (67)) the decision model is fomulated by means of the sequence $(\bar{r}_n)$ of functions $\bar{r}_n := \sum_1^n r_\nu$ instead of the sequence (r_n). That procedure simplifies some formulae but obviously leads (under the same assumptions) to the same model and it seems to be less convenient for the formulation of practical problems.

+) Recently we discovered the possibility to weaken the assumptions (EN) and (EP) in a way that takes into account the specific transition law (q_n); cf. Hinderer (71).

8) One might consider another definition of a (deterministic) plan. Let us call a sequence (g_n) of maps $g_n : H_n \to A$ an (admissible deterministic) H-plan, if

$$g_n(h) \in D_n(h),\ n \in \mathbb{N},\ h \in H_n.$$

To take the action $g_n(h)$ whenever the history h occurs at time n means nothing but to use the plan $f=(f_n)$, defined by

$$(2.17) \qquad f_n(y) := g_n(h_{nf}(y)),\ n \in \mathbb{N},\ y \in S^n,$$

(cp. Aoki (67),p.23). Let Δ' be the set of all H-plans. It is easily seen that the map $g \to f$, defined by (2.17) is surjective, hence nothing is gained by the use of H-plans. We have decided to use in chapter 1 (with the exception of section 9) only plans in the sense of our original definition, since that notion seems to be more natural for application. However, the general theory and some formulae may become a bit simpler when stated in terms of H-plans. This becomes more apparent in section 9 where Δ' is identifidied with a subset of the set Δ^r of randomized plans. In chapter 2 we shall mostly build up the general theory for randomized plans and we shall thereof derive some results for deterministic plans via H-plans. It should be noted that randomized plans are in general not capable of a reduction like that defined by (2.17).

3. The principle of optimality and the optimality equation.

R. Bellman states in his book "Dynamic programming" (57), p.83, the famous *principle of optimality* as follows:

"An optimal policy has the property that whatever the initial state and initial decisions are, the remaining decisions must constitute an optimal policy with regard to the state resulting from the first decision".
(A formal statement of the principle for $\overline{p}$-optimal plans,hence also for optimal plans, is given below in theorem 3.8.)

In the literature the principle plays a dominant but not always clear role. The following remarks are an attempt to clarify some points.

1) As usual, the term "principle" shall probably indicate that we are dealing with a statement which is valid for a large class of problems but whose exact range of validity cannot easily be formulated. This implies however that the principle needs a proof for any well-defined model under consideration. Karlin (55) was the first who called attention to this point.

2) R.Bellman states in his book (57) that "a proof of the principle by contradiction is immediate". This is true for deterministic models (cf.section 8) and simple stochastic models. But already the model considered in this chapter requires a little thought (cf.theorem 3.8 (iii)) and the general case needs the rather heavy measure-theoretic apparatus developped by Blackwell, Strauch and others (cf.chapter II).

3) The importance of the principle does not rest so much on the fact that it furnishes a necessary condition for the optimality of a policy but in the fact that it is often regarded as a convenient tool for deriving the *optimality equation* (OE) (cf.(3.15)), which on its part is the starting point for many investigations in dynamic programming. However, to the best of our knowledge there has never been given a rigorous proof of the OE in the *general case by means of the principle,* though the proofs of the OE and of the principle show some similarities. In remark (iii) after theorem 3.9 we shall give a proof of the OE by means of the principle under rather restrictive assumptions. We also remark that sometimes in the literature the principle and the OE are regarded as the

same statement, though these are definitely two different things.

It might be astonishing that, while a rigorous proof of the OE by means of the principle presents difficulties, one might on the contrary use the OE for a proof of the principle (cf.remark (ii) after the proof of theorem 3.9).

4) If there would exist a proof of the OE by means of the principle it would nevertheless be of limited value, because this would establish the OE only for those cases where a $\bar{p}$-optimal plan exists. For practical purposes it is important that the OE is valid whether there exists a $\bar{p}$-optimal plan or not. Thus it is always possible to compute in principle the supremum (on the set of plans) of the expected total reward and compare it with the expected total reward of a plan which one suspects to be a "good" one.

5) One could think of using the principle as a criterion of optimality. However, it is only a *necessary* condition, while in practice one usually looks for sufficient criteria. In theorem 5.1 the usual sufficient condition is weakened such that it becomes also necessary. The proof of the necessity part uses the principle while the proof of the sufficiency part uses the OE.

6) The principle of optimality and the OE are in their mathematical essence statements on the interchangeability of summation (or integration) and the "operator" supremum. This yields a method of proof different from that one used in this section (cf.e.g.White (69),p.168)[+].

For any plan f there is defined the *conditional* expectation $G_{nf}(y)$ of the *reward* for the time period (n,∞) under the condition that the sequence of previous states is y:

$$(3.1) \qquad G_{nf}(y) := E_f\left[\sum_{i=n}^{\infty} r_{if} \circ \eta_i \mid \eta_n = y\right], n \in \mathbb{N}, y \in S^n, f \in \Delta.$$

(Note that according to our definition of conditional expectation in appendix 3, $G_{nf}(y)$ is also defined if $P_f(\eta_n=y)=0$.)
From (2.7) and (2.8) we conclude that

$$(3.2) \qquad -\infty \le G_{nf} \le \sum_{i=n}^{\infty} \|r_i^+\| < \infty \quad \text{in case (EN)},$$

$$(3.3) \qquad -\infty < -\sum_{i=n}^{\infty} \|r_i^-\| \le G_{nf} \le \infty \quad \text{in case (EP)},$$

$$(3.4) \qquad |G_{nf}| \le \sum_{i=n}^{\infty} \|r_i\| \quad \text{in case (C)}.$$

[+]) A rigorous presentation of that method will be published elsewhere.

Consider some fixed history $h=(s_1,a_1,\ldots,s_n)$ at time n, and put $y:=(s_1,s_2,\ldots,s_n)$. At this stage of the process there remain at our disposal those plans $f\in\Delta$, for which the application of f, together with the sequence y of states, resulted in the history h, i.e. for which $h_{nf}(y)=h$, or otherwise for which $f_\nu(y_\nu)=a_\nu, 1\leq\nu<n$. Let us denote this *set of plans* by $\Delta_n(h), n\in\mathbb{N}, h\in H_n$. $\Delta_n(h)$ does not depend on s_n; in particular $\Delta_1(s)=\Delta$ for any $s\in H_1$. Furthermore $\Delta_{n+1}(h,a,s)\subset\Delta_n(h)$ for $(h,a,s)\in H_{n+1}$.

We shall use the following *convention:* If h and y occur in the same context then y shall denote the sequence of states in h, unless otherwise indicated.

Now we may define, since the sets $\Delta_n(h)$ are non-empty,

$$G_n(h) := \sup_{f\in\Delta_n(h)} G_{nf}(y), \quad n\in\mathbb{N}, h\in H_n,$$

which one denotes as the *maximal conditional expected reward* for period (n,∞) under the condition that the process has run through history h. For ease of reference we state explicitely the defining properties of $G_n(h)$:

a) $G_{nf}(y) \leq G_n(h_{nf}(y))$, $n\in\mathbb{N}$, $y\in S^n$;

b) if $G_n(h)=\infty$ then there exists for any number $N\in\mathbb{R}$ a plan $f\in\Delta_n(h)$ such that $G_{nf}(y)\geq N$;

c) if $G_n(h)<\infty$ then there exists for any number $\varepsilon>0$ a plan $f\in\Delta_n(h)$ such that $G_{nf}(y) \geq G_n(h)-\varepsilon$.

Obviously the inequalities (3.2)-(3.4) remain true if G_{nf} is replaced by G_n.

In the sequel we shall need some elementary properties of the "operator" sup. Some of them are so simple that in the literature they usually are overlooked. Nevertheless they seem to clarify some points in the formal presentation of the theory. In the following four lemmas B and C are arbitrary non-empty sets.

Lemma 3.1 Let $g:B\to C$ and $\varphi:C\to\overline{\mathbb{R}}$ be any maps. Then

$$\sup_{b\in B} \varphi(g(b)) = \sup_{c\in C} \varphi(c),$$

provided that the map g is a map *onto* C (i.e. a surjective map). The simple proof is ommited.

Lemma 3.2. Let Z be an arbitrary non-empty subset of $B\times C$, and let $u: Z \to \overline{\mathbb{R}}$ be an arbitrary map. Then

$$\sup_{(b,c)\in Z} u(b,c) = \sup_{b\in pr(Z,B)} \sup_{c\in Z_b} u(b,c).$$

Here pr(Z,B) denotes the projection of Z into B and Z_b is the b-section of Z (see fig.2).

fig. 2

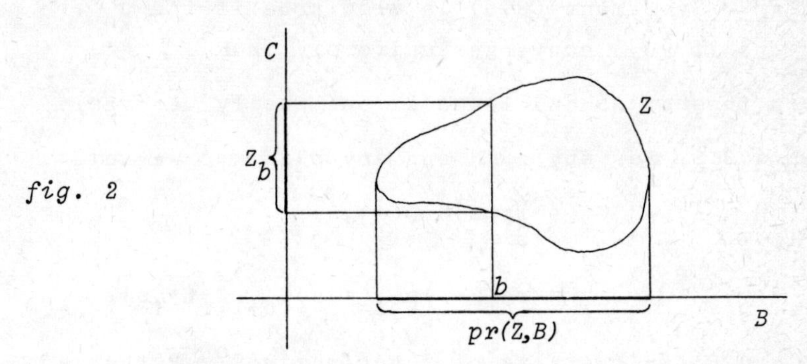

Lemma 3.2 may loosely be stated as: the two-dimensional supremum equals the iterated supremum.

Proof of Lemma 3.2. Denote by ℓ and r the left and right hand side of the equation in the lemma, respectively. For any $(x,y)\in Z$ we get

$$u(x,y) \leq \sup_{a\in Z_x} u(x,a),$$

hence

$$\ell \leq \sup_{(x,y)\in Z} \sup_{a\in Z_x} u(x,a) = r.$$

The inequality $r \leq \ell$ is obtained in the same way. $\lrcorner$

Lemma 3.3. Let the functions $u: B \to \mathbb{R}$ and $v: B \to \mathbb{R}$ be bounded from above.+) Then

$$\left|\sup_b u(b) - \sup_b v(b)\right| \leq \sup_b \left|u(b) - v(b)\right|.$$

Proof. The assertion is trivially true, if sup u = sup v. Hence, let us assume, e.g., sup u < sup v. For any ε such that $0 < \varepsilon \leq \sup v - \sup u$ there exists some $b\in B$ for which

$$u(b) \leq \sup u \leq \sup v - \varepsilon \leq v(b).$$

Therefore $\sup v - \sup u - \varepsilon \leq v(b) - u(b) \leq \sup|v-u|$, which implies $|\sup v - \sup u| = \sup v - \sup u \leq \sup|v-u| + \varepsilon$. Since this is true for all sufficiently small and positive numbers ε the proof is complete. $\lrcorner$

+) It follows that sup u and sup v are finite.

An immediate consequence of lemma 3.2 is the following

Lemma 3.4. Let (u_n) be a non-decreasing sequence of functions $u_n: B \to \overline{\mathbb{R}}$. Then

$$\lim_n \sup_b u_n(b) = \sup_b \lim_n u_n(b). \tag{3.5}$$

Remark. Statement (3.5) is also true if the u_n are real-valued and if (u_n) converges uniformly in $\mathbb{R}$.

From theorem A5 and lemma A7 one easily derives

Lemma 3.5. For any $n \in \mathbb{N}$ and any plan $f \in \Delta$ we have

$$G_f = \sum_s p(s) G_{1f}(s), \tag{3.6}$$

$$G_{nf} = r_{nf} + \sum_s p_{nf}(\cdot, s) G_{n+1,f}(\cdot, s). \tag{3.7}$$

Let B_n^- and B_n^+ be the sets of functions $v: S^n \to \overline{\mathbb{R}}$ that are bounded from above or from below, respectively. For the purpose of simplifying the notation we shall write the right hand side of (3.7) by means of an operator $\Lambda_{nf}: B_{n+1}^- \to B_n^-$ in case (EN) and an operator $\Lambda_{nf}: B_{n+1}^+ \to B_n^+$ in case (EP), respectively. Λ_{nf} is defined by

$$\Lambda_{nf} v := r_{nf} + \sum_s p_{nf}(\cdot, s) v(\cdot, s). \tag{3.8}$$

Obviously Λ_{nf} is isotone (i.e. $v \leq w$ implies $\Lambda_{nf} v \leq \Lambda_{nf} w$) and $\Lambda_{nf}(v+c) = \Lambda_{nf} v + c$ for any constant $c \in \mathbb{R}$. The interpretation of Λ_{nf} is the following: If at time n the sequence y of states $s_1, s_2, \ldots, s_n$ has occurred, if we use plan f and if we receive the *final* reward $v(y,s)$ whenever state s occurs at time n+1, then we can expect to receive $(\Lambda_{nf} v)(y)$ at time n+1.

Formula (3.7) now reads

$$G_{nf} = \Lambda_{nf} G_{n+1,f}, \quad n \in \mathbb{N}, \ f \in \Delta. \tag{3.9}$$

Lemma 3.6. $G_{nf} = \lim_{k \to \infty} \Lambda_{nf} \Lambda_{n+1,f} \cdots \Lambda_{n+k,f} 0$, $n \in \mathbb{N}$, $f \in \Delta$.

Proof. From theorem A4 we get

$$G_{nf}(y) = \sum_{i=n}^{\infty} E_f[r_{if} \circ \eta_i | \eta_n = y] = \lim_k \sum_{i=n}^{n+k} E_f[r_{if} \circ \eta_i | \eta_n = y] =$$

$$= \lim_k E_f[\sum_{i=n}^{n+k} r_{if} \circ \eta_i | \eta_n = y] =: \lim_k E_{nk}(y).$$ Now we have

$E_{no}(y) = r_{nf}(y) = (\Lambda_{nf} 0)(y)$ $\forall n$. Let us assume that

$E_{nk} = \Lambda_{nf} \Lambda_{n+1,f} \cdots \Lambda_{n+k,f} 0$ holds for $n \in \mathbb{N}$ and $k = 0,1,\ldots,N$. Then

$$E_{n,N+1}(y) = E_f[\sum_{i=n}^{n+N+1} r_{if} \circ \eta_i | \eta_n = y] = r_{nf}(y) + \sum_s p_{nf}(y,s) E_{n+1,N}(y,s)$$

$$= \Lambda_{nf}(\Lambda_{n+1,f} \cdots \Lambda_{n+N+1,f} 0).$$ By induction on k the assertion follows. ┘

In the sequel we shall use the set T_{nf} of those points $y \in S^n$ that occur under the use of f with positive probability, i.e.

(3.10) $$T_{nf} := \{y \in S^n : P_f(\eta_n = y) > 0\},$$

or more explicitely

(3.11) $$T_{nf} := \{y \in S^n : \prod_{\nu=1}^{n} p_{\nu-1,f}(y_\nu) > 0\},$$

$$y_\nu := (s_1, s_2, \ldots, s_\nu).$$

T_{nf} is called the *support* of the distribution of η_n under the probability measure P_f. T_{nf} depends only on $f_1, f_2, \ldots, f_{n-1}$, in particular $T_{1f} := \{s \in S : p(s) > 0\} =: T_1$ is independent of f. We have the recursion

(3.12) $$T_{n+1,f} := \{(y,s) \in T_{nf} \times S : p_{nf}(y,s) > 0\}, \ n \in \mathbb{N}.$$

From lemma 3.5 one easily obtains by induction on n:

Lemma 3.7. If $-\infty < G_f < \infty$, then

$$-\infty < G_{nf}(y) < \infty \quad \text{for } n \in \mathbb{N}, y \in T_{nf}.$$

One would expect that the property of a plan f to be $\bar{p}$-optimal or not would depend only on the values of f_n on T_{nf}. In fact, we have

Theorem 3.8. If $-\infty < G < \infty$, then for any plan f the following three statements are equivalent:

(i) f is $\bar{p}$-optimal,

(ii) $G_{1f} = G_1$ on T_1,

(iii) $G_{nf}(y) = G_n(h_{nf}(y)) \quad n \in \mathbb{N}, \ y \in T_{nf}$.

Obviously the implication (ii)$\Rightarrow$(iii) represents for n=2 the *principle of optimality*.

Proof. α) (i)$\Rightarrow$(ii): Assume f to be $\bar{p}$-optimal, but $G_{1f}(i)<G_1(i)$ for some $i\in T_1$, i.e. for some i such that $p(i)>0$. Then there is some plan $g\in\Delta$ for which $G_{1g}(i)>G_{1f}(i)$. Now we define a plan $\varphi=(\varphi_n)$ by deciding to use at time n the function g_n if the process started in i and the function f_n otherwise. Obviously φ is a plan with the properties $G_{1\varphi}(i)=G_{1g}(i)$, $G_{1\varphi}(s)=G_{1f}(s)$ for $s\neq i$. Keeping in mind that $G\in\mathbb{R}$, we get $G=G_f=\sum_s p(s)G_{1f}(s)<\sum_s p(s)G_{1g}(s)=G_g\leq G$ which is a contradiction.

β) (ii)$\Rightarrow$(i): For any $g\in\Delta$ we have $G_f=\sum p(s)G_{1f}(s)= =\sum_{s\in T_1} p(s)G_1(s)\geq\sum_{s\in T_1} p(s)G_{1g}(s)=G_g$, consequently f is $\bar{p}$-optimal.

γ) (ii)$\Rightarrow$(iii): The proof goes by induction on n. Statement (iii) is true for n=1, because this is exactly statement (ii). Now assume (iii) to be true for some $n\in\mathbb{N}$, but

$$(3.13)\qquad G_{n+1,f}(x,j) < G_{n+1}(h_{n+1,f}(x,j)) \quad \text{for some } (x,j)\in T_{n+1,f}.$$

It follows that $G_{n+1,f}(x,j)<G_{n+1,g}(x,j)$ for some plan $g\in\Delta_{n+1}(h_{n+1,f}(x,j))$. Now we construct a plan $\varphi=(\varphi_n)$ by deciding to use g if the state-history at time n+1 is (x,j) and f otherwise. More precisely, we put

$$\varphi_\nu := f_\nu \quad \text{for } 1\leq\nu\leq n$$

$$\varphi_\nu(y_\nu) := \begin{cases} g_\nu(y_\nu) \text{ for } \nu\geq n+1 \text{ and } y_{n+1}=(x,j) \\ f_\nu(y_\nu) \text{ for } \nu\geq n+1 \text{ and } y_{n+1}\neq(x,j). \end{cases}$$

It is easily seen that $\varphi\in\Delta_n(h_{nf}(x))$. From (3.9) and lemma 3.7 we get

$$(3.14)\quad G_{nf}(x)=\Lambda_{nf}G_{n+1,f}(x)<\Lambda_{n\varphi}G_{n+1,\varphi}(x)=G_{n\varphi}(x)\leq G_n(h_{nf}(x))^{+)}$$

which contradicts our assumption that (iii) is true for n. Therefore (iii) is valid for n+1.

δ) (iii)$\Rightarrow$(ii) is trivial. ⌋

Theorem 3.9 (cf.Bellman (57),Dynkin (65) and others). The sequence (G_n) of maximal expected rewards satisfies the

---

+) $\Lambda_{nf}G_{n+1,f}(x)$ denotes the value of the map $\Lambda_{nf}G_{n+1,f}: S^n\to\overline{\mathbb{R}}$ at the point x.

system of optimality equations

$$(3.15)\qquad G_n(h) = \sup_{a\in D_n(h)} \left[r_n(h,a)+\sum_s p_n(h,a,s)G_{n+1}(h,a,s)\right],\quad n\in\mathbb{N}, h\in H_n.$$

We shall write the OE by means of the operator L_n, defined by

$$(3.16)\qquad (L_n w)(h,a) := r_n(h,a)+\sum_s p_n(h,a,s)w(h,a,s).$$

L_{n-1} is defined in case (EN) on the set M_n^- of all functions $w: H_n \to \overline{\mathbb{R}}$, bounded from above, and in case (EP) on the set M_n^+ of all functions $w: H_n \to \overline{\mathbb{R}}$, bounded from below. One easily verifies that for $f\in\Delta_n(h)$

$$(3.17)\qquad (\Lambda_{nf}G_{n+1}\circ h_{n+1,f})(y) = L_nG_{n+1}(h,f_n(y)).\ ^{+)}$$

The interpretation of L_n ist as follows: If at time n the history h has occurred, if we decide to take action a, and if we receive the *final* reward w(h,a,s) whenever state s occurs at time n+1, then we can expect to receive $L_nw(h,a)$ at time n+1.

By means of the operator L_n we may write (3.15) as

$$(3.18)\qquad G_n = \sup_{a\in D_n(\cdot)} L_nG_{n+1}(\cdot,a),\quad n\in\mathbb{N}.$$

Proof of theorem 3.9 (cf. Sirjaev (67)). α) From (3.9) and (3.17) we get

$$\begin{aligned} G_n(h) &= \sup_{f\in\Delta_n(h)} G_{nf}(y) = \sup_f \Lambda_{nf}G_{n+1,f}(y) \leq \\ &\leq \sup_f (\Lambda_{nf}G_{n+1}\circ h_{n+1,f})(y) = \sup_f L_nG_{n+1}(h,f_n(y)) = \\ &= \sup_{a\in D_n(h)} L_nG_{n+1}(h,a). \end{aligned}$$

The validity of the last equality follows from lemma 3.1 since the map that carries $f\in\Delta_n(h)$ to $f_n(y)\in D_n(h)$ is surjective.

β) Now we shall show that the inequality in part α) may be reversed. We fix $(h,a)\in K_n$ and denote the sequence of states in h by x.

Case (i). We suppose that $G_{n+1}(h,a,i)<\infty$ for all i for which $p_n(h,a,i)>0$. For any $\varepsilon>0$ we can find some plan $f^{(i)}\in\Delta_{n+1}(h,a,i)$ such that $G_{n+1,f^{(i)}}(x,i)\geq G_{n+1}(h,a,i)-\varepsilon$. Now we construct the plan φ by defining

+) Here $h_{n+1,f}$ denotes that map from S^{n+1} to H_{n+1}, that assumes the value $h_{n+1,f}(y)$ at the point $y\in S^{n+1}$.

$$\varphi_\nu(y_\nu)=\begin{cases} f_\nu^{(i)}(y_\nu) & \text{for } \nu\geq n+1 \text{ and } y_{n+1}=(x,i),\\ f_\nu(y_\nu), & \text{otherwise,}\end{cases}$$

where $f\in\Delta_{n+1}(h,a,i)$ is an arbitrary plan. (Note that $\Delta_{n+1}(h,a,i)$ does not depend on i.) It is easily seen that $\varphi\in\Delta_n(h)$, $\varphi_n(x)=a$ and $G_{n+1,\varphi}(x,i)=G_{n+1,f^{(i)}}(x,i)\geq G_{n+1}(h,a,i)-\varepsilon$ whenever $p_n(h,a,i)>0$. Now we get

$$G_n(h)\geq G_{n\varphi}(x) = \Lambda_{n\varphi}G_{n+1,\varphi}(x) \geq \Lambda_{n\varphi}G_{n+1}\circ h_{n+1,\varphi}(x)-\varepsilon$$
$$=L_nG_{n+1}(h,a)-\varepsilon.$$

Since $\varepsilon>0$ was arbitrary, we have $G_n(h)\geq L_nG_{n+1}(h,a)$ for every $a\in D_n(h)$. Combining this result with α) we infer that (3.15) holds.

Case (ii): We suppose that $G_{n+1}(h,a,i)=\infty$ for some $i\in S$ for which $p_n(h,a,i)>0$. From (3.2) we realize that this can happen only in case (EP). For any $k\in\mathbb{N}$ we can find some $f^{(k)}\in\Delta_{n+1}(h,a,i)$ such that $G_{n+1,f^{(k)}}(x,i)\geq k$, hence

$$G_{n,f^{(k)}}(x) = r_n(h,a)+\sum_s p_n(h,a,s)G_{n+1,f^{(k)}}(x,s)$$
$$\geq r_n(h,a)+kp_n(h,a,i)-\sum_{n+1}^{\infty}\|r_\nu^-\|.$$

Therefore $G_n(h)=\infty$, and part α) shows that (3.15) is true. $\lrcorner$

Remarks. (i) Let us define in the cases (EN) and (EP) operators $U_n:M^-_{n+1}\to M^-_n$ and $U_n:M^+_{n+1}\to M^+_n$ by means of

$$(3.19)\qquad U_nv_{n+1} := \sup_{a\in D_n(\cdot)} L_nv_{n+1}(\cdot,a), n\in\mathbb{N}.$$

U_n is isoton, i.e. $v\leq w \Rightarrow U_nv\leq U_nw$. Moreover $U_n(v+c)=U_nv+c$ for any constant $c\in\mathbb{R}$. Then (3.15) takes on the form

$$(3.20)\qquad G_n = U_nG_{n+1},\ n\in\mathbb{N}.$$

Finally we may carry notations to extremes by defining the operator U that sends any sequence $v=(v_n)$ of functions $v_n\in M_n^+$ $\forall n$ or $v_n\in M_n^-$ $\forall n$ to the sequence Uv defined by

$$(3.21)\qquad (Uv)_n := U_nv_{n+1}, n\in\mathbb{N}.$$

Then theorem 3.9 tells us that (G_n) satisfies the OE

$$(3.22)\qquad U(G_n) = (G_n),$$

i.e. that the sequence (G_n) is a *fixed point* of the operator U.

(ii) The construction of the plan φ in the proof of the implication (ii)$\Rightarrow$(iii) in theorem 3.8 (which is the essential step in the proof of the principle of optimality) can be circumvented by using the OE (3.18). Indeed, if (3.13) holds then

$$G_{nf}(x) = \Lambda_{nf} G_{n+1,f}(x) < \Lambda_{nf} G_{n+1} \circ h_{n+1,f}(x) \leq$$

$$\leq \sup_{a \in D_n(h_{nf}(x))} L_n G_{n+1}(h_{nf}(x), a).$$

According to the OE the last term equals $G_n(h_{nf}(x))$ which contradicts our assumption that (iii) is true for n.

(iii) We shall here present a proof of the OE by means of the principle of optimality under some restrictive assumptions. Let us assume that there exists a $\bar{p}$-optimal plan f and that $-\infty < G < \infty$. The easy part of the proof of the OE is the inequality

$$G_n \leq \sup_{a \in D_n(\cdot)} L_n G_{n+1}(\cdot, a), \tag{3.23}$$

which is easily derived (cf.part α) of the proof of theorem 3.9). Now assume that in (3.23) the inequality is strict for n=1 and some $i \in S$. Then there exists some $b \in D_1(i)$ such that $G_1(i) < L_1 G_2(i,b)$. In order to construct a contradiction to our assumption one will try to modify f to a plan g by just replacing $f_1(i)$ by $g_1(i) := b$ and try to show that $G_1(i) < G_{1g}(i)$ in the following way: At first we have

$$G_1(i) < r_1(i,b) + \sum_s p_1(i,b,s) G_2(i,b,s). \tag{3.24}$$

If we are able to show that

α) $g \in \Delta$,

β) $G_{2g} = G_{2f}$ whenever $p_1(i,b,s) > 0$,

γ) $(i,s) \in T_{2f}$ whenever $p_1(i,b,s) > 0$,

δ) $G_2(i,b,s) = G_2(i,f_1(i),s)$ whenever $p_1(i,b,s) > 0$

then we get, since $G_{2f}(i,s) = G_2(i,f_1(i),s)$ for any point $(i,s) \in T_{2f}$ by the principle of optimality, the equation $G_2(i,b,s) = G_2(i,f_1(i),s) = G_{2f}(i,s) = G_{2g}(i,s)$. Then $G_1(i) < r_1(i,g(i)) + \sum_s p_{1g}(i,s) G_{2g}(i,s) = G_{1g}(i)$ yields the desired contradiction.

Now let us look at the conditions α)-δ). None of them is necessarily true for all models. However, if we restrict attention to Markovian models (cf.section 4), which are the models usually dealt with in the literature, then D_n and G_n depend only on s_n, f may be assumed Markovian, and therefore α),β) and δ) are true. Condition γ) need not be true; it is true if e.g. $p_1(i,a,s)>0$ for all $(i,a,s)\in H_2$. Therefore there seems to exist a proof of the OE by means of the principle only in rather special cases.

We are going to prove a result that is intimately connected with the OE.

Theorem 3.10. $G = \sum_s p(s)G_1(s)$.

This theorem tells us that $\sup_{f\in\Delta} \sum_s p(s)G_{1f}(s)=\sum_s p(s)\sup_{f\in\Delta} G_{1f}(s)$, i.e. that taking the supremum over $f\in\Delta$ and summation over $s\in S$ are interchangeable operations.

A proof of theorem 3.10 may be given right along the lines of the proof of the OE. (If $-\infty<G<\infty$, and if there exists a $\bar{p}$-optimal plan, then theorem 3.10 is easily proved by means of statement (ii) of theorem 3.8.) Another proof is provided by the following method of "shifting" the original model one unit along the time axis. Define a model $(S,A,D',(p_n'),(r_n'))$ in the following way:
$D_1'(s):=A, s\in S$; p' arbitrary; $r_1':\equiv 0$; and for $n\in\mathbb{N}, (h,a,s)\in H_{n+1}$, $s_1\in S$, $a_1\in A$, we put

$$D'_{n+1}(s_1,a_1,h): = D_n(h),$$

$$p'_{n+1}(s_1,a_1,h,a,s): = p_n(h,a,s),$$

$$r'_{n+1}(s_1,a_1,h,a): = r_n(h,a).$$

Then one can prove that $G=G_1'(x)$ and $G_1(s)=G_2'(x,a,s)$ for any $x\in S, a\in A$. The OE, applied for n=1 to the shifted model, implies immediately theorem 3.10. ⌋

Remark. It is possible to derive the OE from theorem 3.10 by means of a reduction method, which may be useful also in other situations, *and which we call the standard reduction method.* Roughly spoken, the idea consists in treating that part of the DM that happens after the fixed time $t\in\mathbb{N}$ as a new model. More precisely, let us fix some point $k=(h,a)\in K_t$, and let us define a DM $(S,A,D',(p_n'),(r_n'))$ as follows:

$$D_n'(h'): = D_{t+n}(k,h'), h'\in H_n': = (H_{t+n})_k;$$

$$p': = p_t(k,\cdot), p_n'(h',a',\cdot): = p_{t+n}(k,h',a',\cdot), r_n'(h',a'):=$$

$$=r_{t+n}(k,h',a'), n\in\mathbb{N}, h'\in H_n', a'\in D_n'(h').$$

It follows that $H_1^! = S$ and
$H'_{n+1} = \{(h',a',s') \in \overline{H}_{n+1} : h' \in H'_n, a' \in D'_n(h')\}$, hence our definitions are meaningful.

One can verify the following properties:

(i) If case (EN) or (EP) holds for (r_n), then case (EN) or (EP) holds for (r'_n), respectively.

(ii) Denote the sequence of states in the fixed history h by y. Then the equation

$$f'_n(y') := f_{t+n}(y,y'), \; n \in \mathbb{N}, \; y' \in S^n,$$

defines a map $f \to f'$ from $\Delta_{t+1}(k,s)$, s arbitrary, onto Δ'.

(iii) $G'_{1f'} = G_{t+1,f}(y,\cdot)$, $f \in \Delta_{t+1}(k,s)$, and

$$G'_1 = G_{t+1}(k,\cdot). \tag{3.25}$$

Now we shall apply this reduction method for the derivation of the OE from theorem 3.10. At first we have for fixed $t \in \mathbb{N}$, $h \in H_n$, by lemma 3.5 the relation

$$G_t(h) = \sup_{f \in \Delta_t(h)} \left[r_t(h,f_t(y)) + \sum_s p_t(h,f_t(y),s) G_{t+1,f}(y,s)\right].$$

Put $Z := \{(a,f) \in D_t(h) \times \Delta_t(h) : f \in \Delta_{t+1}(h,a,s)\}$, s arbitrary. The map $f \to (f_t(y),f)$ maps $\Delta_t(h)$ onto Z, hence

$$G_t(h) = \sup_{(a,f) \in Z} \left[r_t(h,a) + \sum_s p_t(h,a,s) G_{t+1,f}(y,s)\right]$$

$$= \sup_{a \in D_t(h)} \left[r_t(h,a) + \sup_{f \in \Delta_{t+1}(h,a,s)} \sum_s p_t(h,a,s) G_{t+1,f}(y,s)\right]$$

by lemma 3.1 and lemma 3.2. The application of the reduction method to theorem 3.10 yields

$$\sup_{f \in \Delta_{t+1}(h,a,s)} \sum_s p_t(h,a,s) G_{t+1,f}(y,s) = \sup_{f' \in \Delta'} \sum_s p'(s) G'_{1f}(s) =$$

$= \sum_s p'(s) G'_1(s) = \sum_s p_t(h,a,s) G_{t+1}(h,a,s)$, hence the OE is proved. _|

<u>Definition</u>. A solution of the OE (3.15) is a sequence (v_n) of maps $v_n \in M_n^+, n \in \mathbb{N}$, or $v_n \in M_n^-$, $n \in \mathbb{N}$, in case (EP) or (EN), respectively, such that $v_n = U_n v_{n+1}, n \in \mathbb{N}$.

It is well known that the OE may have more than one solution (cf.e.g. Radner (67),p.24 or Beckmann (68),p.91). Therefore it is an important task to characterize (G_n) among the set of all solutions of (3.15).

Theorem 3.11 (cp. Blackwell (65),(67) and Strauch (66)).

a) In case (EP) the sequence (G_n) is the (termwise) smallest of those solutions (v_n) of the OE that satisfy

$$(3.26) \qquad v_n \geq -\sum_n^\infty \|r_\nu^-\|, \; n \in \mathbb{N}.$$

b) In case (C) the sequence (G_n) is the only solution (v_n) of the OE that satisfies

$$(3.27) \qquad \|v_n\| \leq \sum_n^\infty \|r_\nu\|, \; n \in \mathbb{N}.$$

Remarks. (i) In the general case (EN) no characterization of (G_n) among the solutions of the OE seems to be known.

(ii) Condition (3.26) may also be written as

$$(3.28) \qquad \|v_n^-\| \leq \sum_n^\infty \|r_\nu^-\|, \; n \in \mathbb{N},$$

from which

$$(3.29) \qquad \lim_n \|v_n^-\| = 0$$

is derived. It is easy to show that in case (EP) the sequence (G_n) is even the smallest of those solutions of the OE that satisfy (3.29). In fact, if (v_n) is such a solution, then $v_n = U_n v_{n+1}$ implies

$$-\|v_n^-\| \geq -\|r_n^-\| - \|v_{n+1}^-\| ,$$

hence

$$\|v_n^-\| \leq \sum_n^{n+k} \|r_\nu^-\| + \|v_{n+k+1}^-\| ,$$

hence (3.28) is true.

(iii) The preceding reasoning may easily be modified in order to show that in case (C) the sequence (G_n) is the only solution (v_n) of the OE for which

$$(3.30) \qquad \lim_n \|v_n\| = 0.$$

Proof of theorem 3.11. a) From (3.3) we easily infer that (G_n) satisfies (3.26). Let (v_n) be any solution of the OE that satisfies (3.26). If we put $R_n := \sum_n^\infty \|r_\nu^-\|$, then $R_n \to 0$ $(n \to \infty)$ and $v_n + R_n \geq 0$ $\forall n \in \mathbb{N}$. Now we shall show by induction on k that for $k \in \mathbb{N}_0$

(3.31) $$G_{n,k} \leq v_n + R_{n+k},\ n \in \mathbb{N}\ ,$$

where (G_{nk}) is defined in theorem 4.1 below. At first, (3.31) is true for k=0 since $G_{no} \equiv 0$. Assume that (3.31) is true for k=m and any $n \in \mathbb{N}$. Then, using the OE and the properties of U_n, we get

$$G_{n,m+1} = U_n G_{n+1,m} \leq U_n(v_{n+1} + R_{m+n+1})$$

$$= U_n v_{n+1} + R_{m+n+1} = v_n + R_{m+n+1},\ n \in \mathbb{N}.$$

Hence (3.31) is true for all $k \in \mathbb{N}_o$. Theorem 4.1 below, which is proved without the use of theorem 3.11, implies

$$G_n = \lim_k G_{nk} \leq \underline{\lim}_k (v_n + R_{n+k}) = v_n.$$

b) From (3.4) one easily derives that (G_n) satisfies the inequality (3.27). Now let (V_n) be another solution of the optimality equation, satisfying (3.27). By means of lemma 3.3 we get

$$|G_n(h) - V_n(h)| = |\sup_{a \in D_n(h)} (L_n G_{n+1})(h,a) - \sup_{a \in D_n(h)} (L_n V_{n+1})(h,a)|$$

$$\leq \sup_{a \in D_n(h)} |(L_n G_{n+1})(h,a) - L_n V_{n+1}(h,a)|$$

$$= \sup_{a \in D_n(h)} |\sum_s p_n(h,a,s)[G_{n+1}(h,a,s) - V_{n+1}(h,a,s)]|$$

$$\leq \sup_{a \in D_n(h)} \sum_s p_n(h,a,s)|G_{n+1}(h,a,s) - V_{n+1}(h,a,s)|$$

$$\leq \|G_{n+1} - V_{n+1}\| .$$

Since this inequality is true for any $h \in H_n$, we get

$$\|G_n - V_n\| \leq \|G_{n+1} - V_{n+1}\|.$$

By induction follows for any $k \in \mathbb{N}$

$$\|G_n - V_n\| \leq \|G_{n+1} - V_{n+k}\| \leq \|G_{n+k}\| + \|V_{n+k}\| \leq 2 \sum_{n+k}^{\infty} \|r_i\| \to 0\ (k \to \infty).$$

Therefore $\|G_n - V_n\| = 0$, hence $G_n = V_n$. ⌋

4. Value iteration.

As we shall show in section 5 below, one can construct (at least in principle) $\overline{p}$-optimal plans, if they exist, by means of the functions G_n. A computation of (G_n) by direct solution of the OE under the pertinent constraints is possible only in rare cases. Hence it is important that an iterative solution of the OE is possible,at least in case (EP).

Theorem 4.1. (cp.Beckmann (68),p.46). Define the double sequence $(G_{nk}, n\in \mathbb{N}, k\in\mathbb{N})$ of functions $G_{nk}: H_n \to \overline{\mathbb{R}}$ by means of the system of recursions

(4.1) $$G_{no} :\equiv 0 \;, \; n\in \mathbb{N}, \text{ and}$$

(4.2) $$G_{nk} := U_n G_{n+1,k-1}, \; n\in \mathbb{N},\; k\in \mathbb{N}.$$

(Here U_n is the operator defined by (3.17).) Then

(i) $$G_{nk}(h) = \sup_{f\in\Delta_n(h)} E_f\Big[\sum_{i=n}^{n+k-1} r_{if}\circ\eta_i \,\Big|\, \eta_n = y\Big], \; n,k\in \mathbb{N}, h\in H_n.$$

(ii) In case (EP) we have $G_n = \lim_{k\to\infty} G_{nk}$, $n\in\mathbb{N}$.

Remark.

The computation of G_{nk} proceeds by the diagonal procedure described in fig.3.

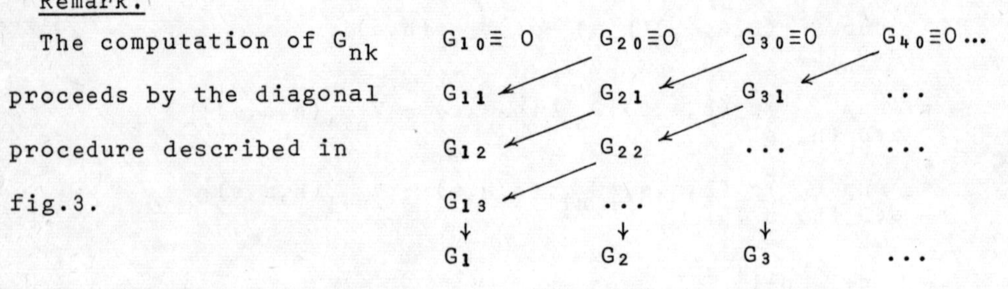

fig.3

Proof. (i) Here one could repeat the proof of the OE, or one can use the OE in the following manner. Let us modify our original decision model DM to a model $DM^{(m)}$, $m\in\mathbb{N}$, by putting $r_\nu^{(m)}$ equal to zero if $\nu\geq m$. If we denote the right hand side of (i) by $V_{nk}(h)$ then we have $V_{nk} = G_n^{(n+k)}$ for all $n\in\mathbb{N}, k\in\mathbb{N}$, and $V_{no}\equiv 0$ for all $n\in\mathbb{N}$. If we fix $n\in\mathbb{N}$ and $k\in\mathbb{N}$, then the OE for $DM^{(n+k)}$ yields

$$V_{nk} = G_n^{(n+k)} = U_n G_{n+1}^{(n+k)} = U_n V_{n+1,k-1}.$$

Hence (V_{nk}) satisfies the system of recursions (4.1) and (4.2), and therefore $(V_{nk}) = (G_{nk})$.

(ii) At first we shall assume that $r_n \geq 0$, $n \in \mathbb{N}$. It follows that $E_f[\sum_{i=n}^{n+k-1} r_{if} \circ \eta_i | \eta_n = y]$ is increasing in k for any fixed f. Hence lemma 3.4 and lemma A2 imply that $V_{nk}(h)$ converges to $\sup_{f \in \Delta_n(h)} E_f[\sum_{i=n}^{\infty} r_{if} \circ \eta_i | \eta_n = y] = G_n(h)$ for $k \to \infty$.

Now we remove the assumption that $r_n \geq 0$ for all n. For that reason we consider a decision model DM' in which r_n is changed to $r_n' := r_n + \|r_n^-\| \geq 0$. Then $G_{nk}' = G_{nk} + \sum_n^{n+k-1} \|r_i^-\|$ and $G_n' = G_n + \sum_n^{\infty} \|r_i^-\|$. We can apply (ii) for the model DM' and get $G_{nk} \to G_n$. ⌋

Now we present an *example*, given by Strauch (66), which shows that part (ii) of theorem 4.1 fails in case (EN). Put $S := \mathbb{N}_o$; $A := \{3,4,\ldots\}$; $D_n(h) := A$; $p(s) := \delta_{s1}$; $p_n(h,a,j) := p(s_n,a,j)$ where

$$p(s,a,j) := \begin{cases} \delta_{oj}, & s=0 \\ \delta_{aj}, & s=1 \\ a^{-1}\delta_{oj} + (1-a^{-1})\delta_{2j}, & s=2 \\ \delta_{s-1,j}, & s>2 ; \end{cases}$$

$r_n(h,a) := -\delta_{3s_n} - a^{-1}\delta_{2s_n}$.

It is not difficult to show by induction on n, that

$G_{nk}(h) = -1_{\{3,4,\ldots,k+2\}}(s_n)$, hence $\lim_k G_{1k}(1) = 0$.

Now let $f = (f_n)$ be a Markov plan. If we start in $s_1 = 1$ and choose $f_n(1) := b \in A$, then we are in the state 3 at time b-1 with probability one. Hence $G_{1f}(1) = E_f[\sum_1^{\infty} r_{\nu f} \circ \eta_\nu | \zeta_1 = 1]$ $\leq E_f[r_{b-1,f} \circ \eta_{b-1} | \zeta_1 = 1] = r_{b-1}(3, f_{b-1}(3)) = -1$. From corollary 9.5 and theorem 18.2 we know, since our DM is Markovian, that $G_1(1)$ equals the supremum of $G_{1f}(1)$ over the set of Markov plans f. Therefore $G_1(1) \leq -1 < \lim_k G_{1k}(1)$.

5. Criteria of optimality and existence of $\bar{p}$-optimal plans.

Theorem 3.8 is a necessary and sufficient criterion for $\bar{p}$-optimality. But it seems to be of rather limited value since its application does not only require the functions G_n (which in principle may be obtained by solving the OE) but also the functions G_{nf} which in general are not easy to obtain. The criterion usually used in the literature reads, generalized to our model, as follows: if $f_n(y)$ is a maximum point of $L_n G_{n+1}(h_{nf}(y),\cdot)$ for all $n \in \mathbb{N}$ and *all* $y \in S^n$ then f is $\bar{p}$-optimal. It is easy to construct counterexamples which show that this criterion is in general not necessary. It becomes necessary if we consider only those $y \in S^n$ that occur under the use of f with positive probability. This leads to

Theorem 5.1 (*criterion of optimality*). We assume case (EN) and $G > -\infty$. The plan f is $\bar{p}$-optimal iff for any $n \in \mathbb{N}, y \in T_{nf}$ the point $f_n(y)$ is a maximum point of the function $L_n G_{n+1}(h_{nf}(y),\cdot)$ (defined on $D_n(h_{nf}(y))$).

Proof. α) Let f be $\bar{p}$-optimal, $n \in \mathbb{N}, y \in T_{nf}$. Then $b := h_{nf}(y) \in H_n$. From theorem 3.8 and the OE we conclude that $L_n G_{n+1}(b, f_n(y)) = \Lambda_n G_{n+1,f}(y) = G_{nf}(y) = G_n(b) = \sup_{a \in D_n(b)} L_n G_{n+1}(b,a)$.
β) Let f satisfy the condition stated in the theorem. The OE implies

$$\Lambda_{nf} G_{n+1} \circ h_{n+1,f}(y) = G_n(h_{nf}(y)), n \in \mathbb{N}, y \in T_{nf}. \tag{5.1}$$

By downward induction we conclude that

$$\Lambda_{1f}\Lambda_{2f}\cdots\Lambda_{nf} G_{n+1} \circ h_{n+1,f}(s) = G_1(s) \text{ for all } s \in T_1. \tag{5.2}$$

From $\Lambda_{nf} G_{n+1} \circ h_{n+1,f}(y) = r_{nf}(y) + \sum_s p_{nf}(y,s) G_{n+1}(h_{nf}(y), f_n(y), s)$

$$\leq r_{nf}(y) + R_n = (\Lambda_{nf} 0)(y) + R_n, \text{ where } R_n := \sum_{n+1}^{\infty} \|r_\nu^+\|,$$

we get

$$G_1(s) \leq (\Lambda_{1f}\cdots\Lambda_{nf} 0)(s) + R_n, \quad s \in T_1, n \in \mathbb{N}.$$

Now lemma 3.6 implies, since $R_n \to 0 (n \to \infty)$, that $G_1(s) \leq G_{1f}(s), s \in T_1$. Theorem 3.8 shows that f must be $\bar{p}$-optimal. ⌋

From theorem 5.1 and the OE we get the

Corollary 5.2. Assume case (EN) and $G > -\infty$. Then the plan f is $\bar{p}$-optimal iff

$$G_n(h_{nf}(y)) = L_n G_{n+1}(h_{nf}(y), f_n(y)) \text{ for all } n \in \mathbb{N}, y \in T_{nf}. \tag{5.3}$$

From theorem 5.1 we easily infer

Theorem 5.3. In case (EN) there exists a $\overline{p}$-optimal plan if for any $n\in\mathbb{N}, h\in H_n$ the function $L_nG_{n+1}(h,\cdot):D_n(h)\to\overline{\mathbb{R}}$ attains its supremum.

Remark. It is easy to show that the condition of theorem 5.3 is not *necessary* for the existence of a $\overline{p}$-optimal plan.

Proof of theorem 5.3. The theorem is true if $G=-\infty$. Assume $G>-\infty$. We define a $\overline{p}$-optimal plan $f=(f_n)$ by induction on n. Let $M_n(h)$ denote the non-empty set of maximum points of $L_nG_{n+1}(h,\cdot)$. At first we define $f_1(s_1)$ to be an arbitrary point in $M_1(s_1)$, if $s_1\in T_1$, and to be an arbitrary point in $D_1(s_1)$ if $s_1\in S-T_1$. Then condition (5.3) is fulfilled for n=1.

Now assume that we have already defined $f_1,f_2,\ldots,f_m$ such that (5.3) holds for $n=1,2,\ldots,m-1$. The set T_{mf} and the points $h_{mf}(y), y\in S^m$, are determined by $f_1,\ldots,f_{m-1}$. Now we define $f_m(y)$ in the following way: $f_m(y)$ is an arbitrary point in $M_m(h_{mf}(y))$, if $y\in T_{mf}$, and an arbitrary point in $D_m(h_{mf}(y))$, otherwise. Then (5.3) holds for n=m. ⌋

Corollary 5.4. If each of the sets $D_n(h), n\in\mathbb{N}, h\in H_n$, is finite, then there exists in case (EN) a $\overline{p}$-optimal plan.

The conclusion of the corollary remains valid, if A is an euclidean d-space, if each of the sets $D_n(h)$ are compact +) and if each of the functions $L_nG_{n+1}(h,\cdot)$ is continuous. It is useful for practical purposes to weaken the condition of continuity to upper semi-continuity.

Definition. Let $(D,\mathcal{T})$ be a topological space with the topology $\mathcal{T}$. A map $W:D\to\overline{\mathbb{R}}$ is called *upper semi-continuous* (abbreviated by u.s.c) if $x,x_n\in D, x_n\to x$ implies $\overline{\lim_n}\, W(x_n)\leq W(x)$.

W is called lower semi-continuous (abbreviated by l.s.c. if -W is u.s.c., i.e. if $x,x_n\in D, x_n\to x$ implies $\underline{\lim_n}\, W(x_n)\geq w(x)$.

Obviously every continuous map is u.s.c. A simple example of a discontinuous u.s.c. map is a function $W:\mathbb{R}\to\mathbb{R}$ which has everywhere one-sided limits $W(x-)$ and $W(x+)$ such that $W(x)\geq\max(W(x+),W(x-))$. Another trivial example is the following:

+) We use 'compact' in the following sense: A subset D of a Hausdorff space $(A,\mathcal{T})$ is called compact, if every cover of D by open sets has a finite subcover.

If S is any countable set then any map $W:S\to\overline{\mathbb{R}}$ is u.s.c. in the discrete topology $\mathfrak{P}(S)$: if $x_n\to x$ in the discrete topology then $x_n=x$ for n large enough, hence $\overline{\lim_n}\, W(x_n)\leq W(x)$ is trivially true.

We summarize some elementary and well-known properties of u.s.c. and l.s.c. functions (cp.e.g. Natanson (61)).

Lemma 5.5 Let $(D,\mathcal{T})$ be a topological space, $f:D\to\overline{\mathbb{R}}$ and $g:D\to\overline{\mathbb{R}}$ arbitrary maps.

(i) If f and g are u.s.c., and if f+g is defined, then f+g is u.s.c.

(ii) If f and g are u.s.c., finite and $f\geq 0, g\geq 0$, then $f\cdot g$ is u.s.c.

(iii) If f is ≥ 0 and l.s.c. and if g is ≤ 0 and u.s.c., then f.g is u.s.c.

(iv) The limit f of a decreasing sequence of u.s.c. functions $f_k\leq 0, k\in\mathbb{N}$, is u.s.c.

Proof. Let $\alpha=(\alpha_n)$ and $\beta=(\beta_n)$ be two sequences in $\overline{\mathbb{R}}$; $\alpha\beta:=(\alpha_n\beta_n)$, $\overline{\alpha}:=\overline{\lim}\,\alpha_n$, $\underline{\alpha}:=\underline{\lim}\,\alpha_n$, $\overline{\beta}:=\overline{\lim}\,\beta_n$,etc.

Part (i) follows from the relation $\overline{\lim}\,(\alpha_n+\beta_n)\leq\overline{\lim}\,\alpha_n+\overline{\lim}\,\beta_n$ valid whenever all terms in the relation are defined.

If $\alpha_n\geq 0, \beta_n\geq 0$ then

(5.4) $$\underline{\lim}\,(\alpha_n\beta_n)\geq\underline{\lim}\,\alpha_n\cdot\underline{\lim}\,\beta_n,$$

and, if (α_n) and (β_n) are moreover bounded, then

(5.5) $$\overline{\lim}\,(\alpha_n\beta_n)\leq\overline{\lim}\,\alpha_n\cdot\overline{\lim}\,\beta_n.$$

Part (ii) follows from (5.5) and part (iii) follows from (5.4), since $-g\geq 0$ is l.s.c., hence $-fg$ is l.s.c., hence fg is u.s.c. Part (iv) follows easily from the relation $f(x_n)\leq f_k(x_n)\leq f_k(x)+\varepsilon$, valid if $x_n\to x$ and if $n>n_o(\varepsilon,k)$. ⌋

The following lemma is well-known.

Lemma 5.6 Any upper semi-continuous map $W:D\to\overline{\mathbb{R}}$ on a compact Hausdorff-space attains its supremum.

The proof is so simple that we reproduce it here. Put $\alpha:=\sup_{x\in D} W(x)$. Then there exist points $x_n\in D$ such that $W(x_n)\to\alpha$ ($n\to\infty$, convergence in $\overline{\mathbb{R}}$). Since D is compact there exists a subsequence (x_{n_k}) of the sequence (x_n) that converges to some point $b \in D$. Since W is u.s.c., we get $W(b) \geq \overline{\lim_n}\, W(x_n) = \alpha$. Therefore W attains its supremum at b. ⌋

Now we get immediately from theorem 5.3

Theorem 5.7. Let A be a Hausdorff space (under some topology), such that each of the sets $D_n(h), n \in \mathbb{N}, h \in H_n$ is compact. If each of the functions $L_n G_{n+1}(h,\cdot) := D_n(h) \to \overline{\mathbb{R}}$ is upper semi-continuous then there exists in case (EN) a $\overline{p}$-optimal plan.

Our aim will be to find sufficient conditions for the upper semi-continuity of $L_n G_{n+1}(h,\cdot)$ which are easily to check. This may be achieved by using ideas of Maitra (68).

For the reason of simpler notation we shall now introduce some notions which, at first sight, seem to be somewhat artificial. For the remainder of this section we assume that A is endowed with a metrizable topology $\mathcal{T}$. As we have already mentioned above, the system $\mathcal{P}(S)$ of all subsets of S is a metrizable topology in S. In the sequel we shall assume that H_n and K_n are endowed with the relative topology of the product topologies defined by the factor topologies $\mathcal{P}(S)$ and $\mathcal{T}$. Since we use in S the discrete topology, a function $w: H_{n+1} \to \overline{\mathbb{R}}$ is u.s.c. iff it is, loosely spoken, jointly u.s.c. in the action variables $a_1, a_2, \ldots, a_n$.

Lemma 5.8 (cp. Maitra (68), lemma 4.1).

If $w: H_{n+1} \to \overline{\mathbb{R}}$ is bounded from above and u.s.c., and if p_n is l.s.c., then

$$v := \sum_s p_n(\cdot,s) w(\cdot,s)$$

is u.s.c.

Proof. At first we assume that $w \leq 0$ holds. Then $p_n(\cdot,s)w(\cdot,s)$ is u.s.c. for any $s \in S$ according to part (iii) of lemma 5.5. If S is finite, then v is u.s.c. according to part (i) of lemma 5.5. If S is infinite, then v is the limit of a decreasing sequence of u.s.c. functions, hence u.s.c. by part (iv) of lemma 5.5. - Now assume w to be bounded from above and u.s.c; then $w - \|w^+\|$ is non-positive and u.s.c., hence $\sum_s p_n(\cdot s)[w(\cdot,s) - \|w^+\|] = v - \|w^+\|$ is u.s.c. which finally implies that v is u.s.c. ⌋

We shall show that in lemma 5.8 one may replace the assumption of lower semi-continuity of p_n by other conditions.

Lemma 5.9. Let us assume that $w:H_{n+1} \to \overline{\mathbb{R}}$ is bounded and u.s.c., that p_n is u.s.c. and that $\sum_s \sup_{k \in K_n} p_n(k,s) < \infty$ for $n \in \mathbb{N}, h \in H_n$. Then

$$v := \sum_s p_n(\cdot,s) w(\cdot,s)$$

is u.s.c.

Proof. At first we assume w to be ≥ 0. Then $p_n w$ is u.s.c. according to lemma 5.5, part (ii). Now we consider a sequence (k_m) in K_n converging to some point $k_0 \in K_n$. We put $\gamma_m := p_n(k_m,\cdot) w(k_m,\cdot), m \in \mathbb{N}_0$. We have $\gamma_m \leq \|w\| \cdot \sup_k p_n (k,\cdot) =: t$ and $\sum_s t(s) < \infty$. Hence the lemma of Fatou holds and results in $\overline{\lim_m} v(k_m) = \overline{\lim_m} \sum_s \gamma_m(s) \leq \sum_s \overline{\lim_m} \gamma_m(s) \leq$ $\leq \sum_s \gamma_0(s) = v(k_0)$. Therefore v is u.s.c. - If w is bounded and u.s.c., then $w + \|w^-\| \geq 0$ is u.s.c. The preceding argument shows that $v + \|w^-\|$ is u.s.c., hence v is u.s.c. ⌋

Lemma 5.10. (cp. Maitra (68), lemma 3.4)

Let A be compact and let K_n be closed in $H_n \times A$. Let $u: K_n \to \mathbb{R}$ be u.s.c. Then $t := \sup_{a \in D_n(\cdot)} u(\cdot,a)$ is u.s.c..

Proof. Let h be an arbitrary point in H_n. Obviously the set $D_n(h) = (K_n)_h$ is closed and hence compact, since A is compact. The map $u(h,\cdot)$ is u.s.c. on $D_n(h)$, since u is u.s.c. on K_n. Hence there exists a maximum point of $u(h,\cdot)$. Now let (h_m) be a sequence of points in H_n that converges to h. Select a maximum point a_m of $u(h_m,\cdot)$. There exists a subsequence (m') of $\mathbb{N}$ such that $t(h_{m'})$ converges to $\overline{\lim_m} t(h_m)$. Since A is compact, there exists a subsequence (m'') of (m') such that $(a_{m''})$ converges to some point $a \in A$. Then $((h_{m''}, a_{m''}))$ is a sequence of points in K_n that converges to $(h,a) \in H_n \times A$. Since K_n is closed, we have $(h,a) \in K_n$, hence $a \in D_n(h)$. Finally the u.s.c. property of u implies

$$\overline{\lim_m} t(h_m) = \lim_{m''} t(h_{m''}) =$$

$$= \lim_{m''} u(h_{m''}, a_{m''}) \leq u(h,a) \leq t(h).$$

Therefore t is u.s.c. ⌋

Theorem 5.11. We make the following assumptions for any $n \in \mathbb{N}$.

(i) A is a metric compact space;

(ii) the sets $K_n \subset H_n \times A$ are closed;

(iii) p_n is l.s.c.;

(iv) the functions r_n are u.s.c.;

(v) case (C) holds.

Then the functions G_n are u.s.c. and there exists a $\bar{p}$-optimal plan.

Proof. We fix $n \in \mathbb{N}$. We can assume that $r_i \leq 0$, for otherwise we can replace r_i by $r_i - \|r_i^+\| \leq 0$. From theorem 4.1 we know that $G_n = \lim_k G_{nk}$, where G_{nk} is defined by the recursion (4.1) and (4.2), which shows that

$$(5.6) \qquad G_{nk} = U_n U_{n+1} \cdots U_{n+k-1} 0.$$

Let $w: H_{n+1} \to \mathbb{R}$ be an arbitrary u.s.c. function. From lemma 5.8, part (i) of lemma 5.5 and lemma 5.10 we infer that $U_n w$ is u.s.c. Downward induction on k in (5.6) shows that G_{nk} is u.s.c. for any $k \in \mathbb{N}$. Our assumption $r_i \leq 0$ implies together with part (i) of theorem 4.1 that G_{nk} is decreasing in k. Part (iv) of lemma 5.5 tells us that G_n is u.s.c. The existence of a $\bar{p}$-optimal plan is now a consequence of lemma 5.8 and theorem 5.7.

Remark. According to lemma 5.9 one may replace assumption (iii) in theorem 5.11 by the assumption

(iii') $\quad p_n$ is u.s.c. and $\sum_s \sup_{k \in K_n} p_n(k,s) < \infty, n \in \mathbb{N}$.

6. Sufficient statistics, Markovian and stationary models.

In our general model of section 2 we have admitted that D_n, p_n, r_n uses all of the history h at time n. However, a numerical solution is only possible, if the dependence on h is of some special kind, i.e. if the dependence on h holds only by means of some 'simple' function $t_n(h), h \in H_n$. The most important example is the case $t_n(h) := s_n$, which leads to the extensively studied Markovian models. Another important example is provided by the models under uncertainty (cf. section 10), where $t_n(h)$ consists of s_n and the a posteriori distribution of the unknown transition law.

The concept of a sufficient statistic is in wide use in mathematical statistics. We shall use here a somewhat different notion which seems to be more easily applicable for DM's. In the remark after theorem 6.1 we relate our notion of sufficiency to the notion of a sufficient partition, introduced by Dynkin (65).++)

If we want to have that the functions G_n depend on h only by means of $t_n(h)$, then we have to assume some relation between t_{n+1} and t_n. More precisely we use the

<u>Definition</u>. Let F_n be an arbitrary non-empty set and $t_n: H_n \to F_n$ an arbitrary surjective map. +)
The sequence (t_n) is called a *sufficient statistic* of the DM $(S,A,D,(p_n),(r_n))$ if it has the following properties:

$$(\alpha) \quad \left\{ \begin{array}{l} D_n(h) = D_n'(t_n(h)), \\ p_n(h,a,s) = p_n'(t_n(h),a,s), \\ r_n(h,a) = r_n'(t_n(h),a), \\ \text{for some functions } D_n', p_n', r_n'. \end{array} \right\} \quad n \in \mathbb{N},\ (h,a,s) \in H_{n+1}.$$

(β) If $n \in \mathbb{N}, h \in H_n, h' \in H_n$, then $t_n(h) = t_n(h')$ implies $t_{n+1}(h,a,s) = t_{n+1}(h',a,s)$ for all $a \in D_n(h)$ and $s \in S$.

There exists always the trivial sufficient statistic $t_n(h) := h$, $n \in \mathbb{N}, h \in H_n$. Of course, one looks for sufficient

---

+) This is no restriction, since always F_n may be replaced by $t_n(H_n)$.

++) More on the role of sufficient statistics in dynamic programming can be found in Sirjaev (67).

statistics where t_n "contracts" the set H_n substantially. Condition (β) implies the existence of a unique function T_n, defined on the set $\{(t,a):t\in F_n, a\in D_n'(t)\}\times S$ and taking values in F_{n+1}, such that

$$t_{n+1}(h,a,s) = T_n(t_n(h),a,s), n\in\mathbb{N},\ (h,a,s)\in H_{n+1}.$$

Let Q_n^+ and Q_n^- be the set of maps $v:F_n\to\overline{\mathbb{R}}$, that are bounded from below or from above, respectively. The operator U_n', defined by

$$U_n'v:= \sup_{a\in D_n'(\cdot)} [r_n'(\cdot,a)+\sum_j p_n'(\cdot,a,j)v(T_n(\cdot,a,j))],$$

maps Q_{n+1}^+ into Q_n^+ (in case (EP)) and Q_{n+1}^- into Q_n^- (in case (EN)).

Definition. Let (t_n) be a sufficient statistic. A solution of the *reduced optimality equation* is a sequence (v_n) of maps $v_n\in Q_n^+$ (in case (E)) or $v_n\in Q_n^-$ (in case (EN)) such that

$$v_n = U_n'v_{n+1},\ n\in\mathbb{N}.$$

Theorem 6.0. (cp. Dynkin (65)).

Let (t_n) be a sufficient statistic. Then the following statements hold:

(i) There exists for any $n\in\mathbb{N}$ a unique map $G_n':F\to\overline{\mathbb{R}}$ such that

$$G_n = G_n' \circ t_n.$$

(ii) (G_n') is a solution of the reduced optimality equation.

(iii) In case (EP) we have $G_n'=\lim_k U_n'U_{n+1}'\dots U_{n+k-1}'0,\ n\in\mathbb{N}$, and the sequence (G_n') is the termwise smallest of those solutions (v_n) of the reduced OE that satisfy

$$\lim_n \|v_n^-\| = 0. \tag{6.0}$$

(iv) In case (C) the sequence (G_n') is the unique solution (v_n) of the reduced OE for which

$$\lim_n \|v_n\| = 0.$$

Proof. We shall give here a proof only for case (EP). A different proof for the general case may be derived from theorem 18.4 and corollary 9.5. - Consider the map G_{nk}

defined in theorem 4.1 and the map $G'_{nk} := U'_n U'_{n+1} \ldots U'_{n+k-1} 0$. Using the relation $t_{n+1}(h,a,s) = T_n(t_n(h),a,s)$, one easily verifies by induction on k, that $G'_{nk} \circ t_n = G_{nk}$. Keeping in mind that t_n is surjective, we conclude from theorem 4.1 that (i) holds and that $G'_n = \lim_n G'_{nk}$. Part (i) and the OE (3.15) imply that (G'_n) is a solution that satisfies (6.0), since (G_n) does. Finally the characterization of (G'_n) among the solutions of the reduced OE is derived from theorem 3.11 by means of part (i).

Remark 1. Let us define a relation '$\approx$' on $\bigcup_n H_n$ by the definition

$$h \approx h' \iff \begin{cases} h \in H_m \iff h' \in H_m \\ h \in H_m \implies \begin{cases} D_m(h) = D_m(h'), p_m(h,a,s) = p_m(h',a,s), \\ r_m(h,a) = r_m(h',a) \quad \forall a \in D_m(h), s \in S. \end{cases} \end{cases}$$

Moreover, any sufficient statistic $t = (t_n)$ defines a relation $\sim$ on $\cup H_n$ by the definition

$$(6.1) \qquad h \sim h' \iff \begin{cases} h \in H_m \iff h' \in H_m \\ h \in H_m \implies t_m(h) = t_m(h'). \end{cases}$$

It is easily seen that $\sim$ is an equivalence relation with the property

$$(6.2) \qquad h \sim h' \implies h \approx h' \text{ and } (h,a,s) \sim (h',a,s) \quad \forall a \in D_m(h), \; s \in S, \; \text{if } h \in H_m.$$

On the other hand, let $\sim$ be an *arbitrary* equivalence relation on $\cup H_n$ which satisfies (6.2). Then one can prove the existence of a sufficient statistic $(\tilde{t}_n)$ which induces the given relation by means of (6.1).

Dynkin (65) shows for his model and for case (C) a result which reads, adapted to our more general model, as follows: Let $\sim$ be an equivalence relation on $\cup H_n$ which satisfies (6.2), and let $(\tilde{t}_n)$ be an inducing sufficient statistic. Then part (i) of theorem 6.0 is true and there exists for any $\varepsilon > 0$ a strictly ε-optimal plan f such that

$$(6.3) \qquad \tilde{t}_n(h_{nf}(y)) = \tilde{t}_n(h_{nf}(y')) \implies f_n(y) = f_n(y').$$

Dynkin (65) calls then the partition of H_n induced by the relation $\sim$ a *sufficient* partition.

A maximal reduction by means of a sufficient statistic is achieved by the so-called *fundamental partition* introduced by Dynkin (65) and defined by the relation

$$h \overset{*}{\approx} h' \Longleftrightarrow \begin{cases} h\approx h' \text{ and, if } h\in H_m \text{ then } (h,a_m,s_{m+1},\dots,s_k)\approx \\ \approx(h',a_m,s_{m+1},\dots,s_k), \text{ whenever } k>m \text{ and} \\ (h,a_m,s_{m+1},\dots,s_k)\in H_k. \end{cases}$$

It is easily seen that the fundamental partition is the finest of all partitions that satisfy (6.2). Hence, if (t_n^*) is a sufficient statistic that induces the fundamental partition and if (t_n) is an arbitrary sufficient statistic then (t_n) is a function of (t_n^*).

Another reduction which considers the minimal number of effective coordinates in the history h, the so-called minimal memory, has been studied by Hinderer (67). Obviously Markovian models are models for which the length of the minimal memory is not greater than one.

Remark 2. Let us turn to the following question: Let (t_n) be a sufficient statistic. Does there exist to any plan f a plan g such that $G_g \geq G_f$ and such that g_n depends on y only 'by means of t_n' in the sense of property (6.3)? One can verify this statement+) using ideas of Blackwell (65) and Strauch (66), in case (EN) and also in case (EP), if g is allowed to be randomized (cf.section 9). In this section we shall consider in theorem 6.2 below as an application of the optimality criterion 5.2 only the most important special case of the problem.

We have already mentioned the relevancy of so-called Markovian models for which $t_n(h):=s_n$, $n\in\mathbb{N}, h\in H_n$, is a sufficient statistic. In other words, we have the following

Definition. The model $(S,A,D,(p_n),(r_n)$ is called a (general or non-stationary) *Markovian model* if for any $n\in\mathbb{N}$ the sets D_n and the functions p_n and r_n do not depend on $(s_1,a_1,\dots,a_{n-1})$, i.e. if $D_n(h)=D_n(s_n)$ ++), $p_n(h,a,s)= =p_n(s_n,a,s)$, and $r_n(h,a)=r_n(s_n,a), n\in\mathbb{N}$, $(h,a,s)\in H_{n+1}$, holds.

+) cf.section 18

++) Since no confusion is possible, we shall use here and in similar situations this (somewhat ambiguous) terminology.

$D_n(s)$ is now defined for $s \in S$, while $p_n(s,a,\cdot)$ and $r_n(s,a)$ are defined for $s \in S$ and $a \in D_n(s)$. If f is a Markov plan, then (ζ_n) forms an in general inhomogeneous Markov chain on $(\Omega, \mathcal{F}, P_f)$.

In the literature most often special Markovian models, so called stationary models have been studied, which will be dealt with below. The general Markovian model with a special form of the reward functions has recently be investigated by Furukawa (68). In this section we shall derive some properties of Markovian and stationary models from our results previously obtained for general models.

In the remainder of this section we write U_n and G_n instead of U_n' and G_n'.

An immediate consequence of theorem 6.0. is

Theorem 6.1. (cp.Furukawa (68)) Let the model be Markovian.

(i) The maximal expected reward $G_n : H_n \rightarrow \overline{\mathbb{R}}$ depends at most on s_n, i.e. $G_n(h) = G_n(s_n), n \in \mathbb{N}, h \in H_n$.

(ii) The sequence (G_n) satisfies the (reduced) OE

$$G_n(s) = \sup_{a \in D_n(s)} \left[r_n(s,a) + \sum_j p_n(s,a,j) G_{n+1}(j)\right],$$

$$\equiv \sup_{a \in D_n(s)} L_n G_{n+1}(s,a)$$

$$\equiv U_n G_{n+1}(s), n \in \mathbb{N}, s \in S.$$

Let T_{nf}' be the *support of the distribution of* ζ_n *under* P_f, i.e.

$$T_{nf}' := \{s \in S : P_f(\zeta_n = s) > 0\}.$$

Theorem 6.2. (cp. Furukawa (68)). Let the model be Markovian. Assume case (C). +) If there exists a $\overline{p}$-optimal plan, then there exists a Markov $\overline{p}$-optimal plan.

+) It follows from corollary 18.3 that theorem 6.2 holds also in case (EN), and - if randomized plans are allowed - also in case (EP).

Proof. Let f be a $\bar{p}$-optimal plan. According to theorem 6.1 and the necessity part of the optimality criterion given in corollary 5.2 we know that

$$G_n(s_n) = L_n G_{n+1}(s_n, f_n(y)), n \in \mathbb{N}, y \in T_{nf}.$$

According to the sufficiency part of corollary 5.2 we have to construct a Markov plan g such that

$$(6.4) \qquad G_n(s_n) = L_n G_{n+1}(s_n, g_n(s_n)), n \in \mathbb{N}, s \in T'_{ng}.$$

(Note that $y \in T_{ng}$ implies $s_n \in T'_{ng}$.)

We begin the construction of g by defining $g_1 := f_1$. Assume that we have already defined $g_1, \ldots, g_n$. We shall show below that one can then find maps $\varphi_{nk}: S \to S$, $1 \le k \le n$, that satisfy

$$(6.5) \qquad (\varphi_{n1}(s), \varphi_{n2}(s), \ldots, \varphi_{nn}(s), s) \in T_{n+1,f}, \quad s \in T'_{n+1,g}.$$

(Note that $T_{n+1,g}$ is already determined by $g_1, g_2, \ldots, g_n$.)

Then we define

$$(6.6) \qquad g_{n+1}(s) := f_{n+1}((\varphi_{n1}(s), \varphi_{n2}(s), \ldots, \varphi_{nn}(s), s)), s \in S.$$

The plan g which is obtained by the preceeding inductive definition is Markovian and obviously satisfies (6.4).

The inductive construction of the maps φ_{nk} is carried out in the following manner. We start with n=1 and consider an arbitrary point $s \in T'_{2g}$, i.e. a point s for which $\sum_i p(i) p_{1g}(i,s) > 0$. Hence there exists some point $s_1 =: \varphi_{11}(s)$ such that $p(s_1) p_{1g}(s_1, s) > 0$. The last term equals $p(s_1) p_{1f}(s_1, s)$ since $g_1 = f_1$. Therefore $(\varphi_{11}(s), s) = (s_1, s) \in T_{2f}$, and (6.5) is satisfied for n=1. If $P_g(\zeta_2 = s) = 0$, then we define $\varphi_{11}(s)$ arbitrary. Now assume that we have defined φ_{nk} for $1 \le n < m$ and $1 \le k \le n$ in such a way, that (6.5) holds for $1 \le n < m$. Then also $g_1, g_2, \ldots, g_m$ is determined by (6.6). If $s \in T'_{m+1,g}$, i.e. $\sum_i P_g(\zeta_m = i) p_{mg}(i,s) > 0$, then there is a point $s_m =: t(s) \in T'_{mg}$ for which

$$(6.7) \qquad p_{mg}(t(s), s) > 0.$$

Now we define for $1 \le k < m$

$$\varphi_{mk}(s) := \begin{cases} \varphi_{m-1,k}(t(s)), & s \in T'_{m+1,g}, \\ \text{arbitrary}, & \text{otherwise}, \end{cases}$$

Now (6.7) is satisfied for n=m: If $s\in T'_{m+1,g}$, then $t(s)\in T'_{m,g}$ and the induction hypothesis implies that

$$z(s):=(\varphi_{m1}(s),\dots,\varphi_{m,m}(s))=(\varphi_{m-1,1}(t(s)),\dots,t(s))\in T_{mf}.$$

From (6.9) and the definition of g_m follows

$$\begin{aligned} 0<p_{mg}(t(s),s) &= p_m(t(s),g_m(t(s)),s) \\ &= p_m(\varphi_{mm}(s),f_m(z(s)),s) \\ &= p_{mf}(\varphi_{mm}(s),s). \end{aligned}$$

Therefore $(z(s),s)\in T_{m+1,f}$ which verifies (6.5) for n=m. _|

The results of sections 3,4 and 5 yield some *results for Markovian models*. Let us mention the following ones.

Theorem 6.3 (cf.theorem 3.8). Assume $-\infty<G<\infty$.
The Markov plan f is $\overline{p}$-optimal iff $G_{nf}(s)=G_n(s)$ for all $n\in\mathbb{N}$ and $s\in T'_{nf}$.

Theorem 6.4 (cf.theorem 5.1 and corollary 5.2).
Assume case (EN) and $G>-\infty$. Then for any Markov plan f the following statements are equivalent.

α) f is $\overline{p}$-optimal;

β) $f_n(s)$ is a maximum point of the map

$L_nG_{n+1}(s,\cdot):D_n(s)\to\mathbb{R}$, $n\in\mathbb{N}$, $s\in T'_{nf}$;

γ) $G_n(s) = L_nG_{n+1}(s,f_n(s))$, $n\in\mathbb{N}$, $s\in T'_{nf}$.

Theorem 6.5 (cf.theorems 5.11 and 6.2). Let us assume:

α) A is a metric compact space;

β) the sets $D_n(s)$ are closed;

γ) $p_n(s,\cdot,j)$ is l.s.c;

δ) $r_n(s,\cdot)$ is u.s.c;

ε) case (C) holds.

Then there exists for any p a $\overline{p}$-optimal Markov plan.

Blackwell (65) and Furukawa (68) show for their models that a Markov plan f is optimal iff (G_{nf}) satisfies the OE. The situation is somewhat different for $\overline{p}$-optimality, as we shall prove now.

Theorem 6.6. Let the model be Markovian, and assume $-\infty<G<\infty$.

(i) If the Markov plan f is $\overline{p}$-optimal, then (G_{nf}) satisfies the OE, more precisely

(6.8) $\qquad G_{nf}(s) = U_n G_{n+1,f}(s), n\in\mathbb{N}, s\in T'_{nf}.$

(ii) Condition (6.8) is in general *not* sufficient for the $\bar{p}$-optimality of f.

Proof. (i) From (3.7) we get

$$G_{nf}(s) = r_n(s,f_n(s))+\sum_j p_n(s,f_n(s),j)G_{n+1,f}(j) \leq$$
$$\leq \sup_{a\in D_n(s)} [r_n(s,a)+\sum_j p_n(s,a,j)G_{n+1,f}(j)]=U_n G_{n+1,f}(s).$$

On the other hand, we have $G_{n+1,f} \leq G_{n+1}$, therefore $U_n G_{n+1,f} \leq U_n G_{n+1}$. From the OE and from theorem 3.8 we get $U_n G_{n+1}(s)=G_n(s)=G_{nf}(s)$ for all $s\in T'_{nf}$.

(ii) We give the following *counterexample*. Take $S=A=\{1,2\}$, $D_n(s)=A$; $p(1)=1$;

$p_n(i,a,j)=p^a_{ij}$, where $(p^1_{ij}) = \begin{pmatrix}1 & 0\\ 1 & 0\end{pmatrix}$,

$(p^2_{ij}) = \begin{pmatrix}0.5 & 0.5\\ 1 & 0\end{pmatrix}$; $r_1(i,a)=r_2(i,a)=r_{ia}$, where $(r_{ia}) = \begin{pmatrix}1 & 0.5\\ 0 & 4\end{pmatrix}$,

$r_n\equiv 0$ for $n>2$.

Now we consider the Markov plan f for which $f_n\equiv 1$. Condition (6.8) is trivially true for $n>2$, since $G_{nf}\equiv 0$ for $n>2$. We have $T'_{1f}=\{1\}=T'_{2f}$. From the OE we compute for the 'vector' $G_2=\binom{1}{4}$ and likewise $G_1(1)=3$. Moreover $G_{2f}=\binom{r_{11}}{r_{12}}=\binom{1}{0}$ and $G_{1f}(1)=2$. Now it is easily verified that condition (6.8) is satisfied also for n=1 and n=2. However, f is not $\bar{p}$-optimal since $G=G_1(1)=3>2=G_{1f}(1)=G_f$.┘

<u>Definition</u>. The Markovian model $(S,A,D,(p_n),(r_n))$ is called *stationary* if D_n and $p_n (n\in\mathbb{N})$ do not depend on n, and if r_n is of the form $r_n=\beta^{n-1}r$ for some function r and some $\beta\in(0,1\rangle$.

<u>Remarks.</u> 1) The map r is defined on $\{(s,a):s\in S, a\in D(s)\}$.
2) In general, β may be interpreted as discount rate, but some processes with random termination have the same structure (cf. Howard (60/65), p.74, and example 5 in section 8 below). We exclude the trivial case β=0.
3) Most extensively investigated is the case where β<1 and r is bounded (cf. e.g. Howard (60/65), Blackwell (65)). We shall call this case the *discounted case*.
4) For stationary models we have the following equivalences.

Case (EN): either '$r \leq 0$' or '$\beta<1$, r bounded from above';
Case (EP): either '$r \geq 0$' or '$\beta<1$, r bounded from below';
Case (C) = discounted case: $\beta<1$ and r bounded.

5) Blackwell (62) and (65) and Maitra (68) studied the discounted case; Strauch (66) studied the so-called negative case (N) (contained in case (EN)) where $\beta=1$ and $-\infty<r\leq 0$ and the so-called positive bounded case (P) (contained in case (EP)) where $\beta=1$ and r is bounded. The case (P) is also studied by Blackwell (67).

One will expect that in stationary models one is allowed to restrict attention to so-called *stationary plans*, i.e. Markov plans (f_n) where f_n is independent of n. If no confusion is possible, we shall use the symbol f also for the sequence (f_n) for which $f_n=f$.

Theorem 6.7 (cf.Blackwell (65) and Strauch (66).
Let the model be stationary. Then we have:

(i) The sequence (G_n) of maximal expected rewards satisfies

(6.9) $$G_n(h) = \beta^{n-1} G_1(s_n),\ n \in \mathbb{N},\ h \in H_n.$$

(ii) G_1 satisfies the OE

(6.10) $$G_1 = \sup_{a \in D(\cdot)} \left[r(\cdot,a)+\beta\sum_j p(\cdot,a,j)G_1(j)\right] \equiv \sup_{a \in D(\cdot)} L_1\beta G_1(\cdot,a) \equiv U_1\beta G_1.$$

(iii) If $r\geq 0$, then G_1 is the smallest positive solution of (6.10). If r is bounded from below and $\beta<1$, then G_1 is the smallest of those solutions v of (6.10) for which $v \geq -\|r^-\|(1-\beta)^{-1}$. If r is bounded and $\beta<1$, then G_1 is the unique bounded solution of (6.10).

(iv) Define the sequence $(v_k, k \in \mathbb{N}_o)$ recursively by

$$v_o :\equiv 0,$$
$$v_k := U_1(\beta v_{k-1}),\ k \in \mathbb{N}.$$

Then $v_k \to G_1$ $(k\to\infty)$.

Proof. We give here a proof only for case (EP). The general case is settled by theorem 18.6 and corollary 9.5.

(i) From $U_n v = \sup_{a \in D(\cdot)} [\beta^{n-1} r(\cdot,a) + \sum_j p(\cdot,a,j)v(j)]$ we conclude that

$$U_n v = \beta^k U_{n-k}(\beta^{-k} v), \quad 1 \leq k < n < \infty. \tag{6.11}$$

Now it follows by induction on k that $G_{nk} = \beta^{n-1} U_1 U_2 \ldots U_k 0 = \beta^{n-1} G_{1k}$. Part (iii) of theorem 6.0 tells us that $G_{nk} \to G_n$ $(k \to \infty)$ which proves (6.9). Part (ii) is now an immediate consequence of part (i) and of part (ii) of theorem 6.1.
Part (iii) is proved by means of (i) and the parts (iii) and (iv) of theorem 6.0, whereas part (iv) follows from (i) and part (iii) of theorem 6.0. ⌋

For any stationary plan f we have $\Lambda_{nf} v = \beta^k \Lambda_{n-k,f}(\beta^{-k} v)$, $1 \leq k < n < \infty$, hence $G_{nf} = \beta^{n-1} G_{1f}$ by lemma 3.6.

From theorem 6.7 and the results of previous sections, applied earlier to Markovian models we get e.g. the following *results for stationary models*. We define $T_f := \bigcup_n T'_{nf} = \bigcup_n \{s : P_f(\zeta_n = s) > 0\}$. We remark that $T_f = S$ if $p(s) > 0$ $\forall s \in S$.

Theorem 6.8 (cf.theorems 6.3 and 6.6). Let the model be stationary and assume $-\infty < G < \infty$. Then we have

(i) the stationary plan f is $\overline{p}$-optimal iff $G_{1f}(s) = G_1(s)$ for all $s \in T_f$;

(ii) a necessary but not sufficient condition for the $\overline{p}$-optimality of the stationary plan f is that
$G_{1f}(s) = \sup_{a \in D(s)} [r(s,a) + \beta \sum_j p(s,a,j) G_{1f}(j)]$, $s \in T_f$.

Theorem 6.9 (cf.theorem 6.4). Assume $G > -\infty$ and either "$r \leq 0$" or "r bounded from above and $\beta < 1$". Then for any stationary plan f the following statements are equivalent.

α) f is $\overline{p}$-optimal;

β) f(s) is a maximum point of the map
$a \to r(s,a) + \beta \sum_j p(s,a,j) G_1(s)$ $\quad \forall s \in T_f$.

γ) $G_1(s) = r(s,f(s)) + \beta \sum_j p(s,f(s),j) G_1(s)$ $\quad \forall s \in T_f$.

Theorem 6.10 (cf.Blackwell (65) and Strauch (66). Assume $G > -\infty$ and either "$r \leq 0$" or "r bounded from above and $\beta < 1$". If there exists a $\overline{p}$-optimal plan then there exists a stationary $\overline{p}$-optimal plan.

Proof. According to theorems 6.2., 6.4 and 6.7 there exists a $\overline{p}$-optimal Markov plan g which satisfies (6.4), i.e.

$$G_1(s) = L_1\beta G_1(s,g_n(s)),\ n\in\mathbb{N},\ s\in T'_{g_n}.$$

According to theorem 6.9 we have to construct a stationary plan f for which

$$(6.12)\qquad G_1(s) = L_1\beta G_1(s,f(s)),\ s\in T_f.$$

At first we notice that

$$T'_{n+1,g} = \{s\in S: p(i,g_n(i),s)>0 \text{ for some } i\in T'_{ng}\}.$$

For the moment we shall use the abbreviation $\Gamma_n := T'_{ng}$. Now we define recursively a sequence (B_n) of sets $B_n\subset S$ by

$$(6.13)\qquad \begin{aligned} B_1 &:= \Gamma_1 = T_1,\\ B_{n+1} &:= \{s\in S: p(i,g_n(i),s)>0 \text{ for some } i\in B_n\} - \bigcup_1^n B_n,\ n\in\mathbb{N}. \end{aligned}$$

It follows directly from the definition that the sets B_n are pairwise disjoint. Now we define a stationary plan f by

$$(6.14)\qquad f(s) := \begin{cases} g_n(s), & \text{if } n\in\mathbb{N}, s\in B_n,\\ \text{arbitrary point of } D(s), & \text{otherwise.}\end{cases}$$

That plan f will satisfy (6.12) since we are going to prove that

$$(6.15)\qquad B_n\subset\Gamma_n,\ n\in\mathbb{N},$$

and

$$(6.16)\qquad T_f = \sum B_n.$$

Relation (6.15) is true for n=1 by definition of B_1. Now assume that (6.15) is true for some n. For any $s\in B_{n+1}$ there exists some $i\in B_n\subset\Gamma_n$ such that $p(i,g_n(i),s)>0$, hence $s\in\Gamma_{n+1}$. This proves (6.15). We denote T'_{nf} by Φ_n. Equation (6.19) is true if we can show $\bigcup_1^n\Phi_\nu = \sum_1^n B_\nu$ for all $n\in\mathbb{N}$, or equivalently, if

$$(6.17)\qquad \Phi_n - \bigcup_1^{n-1}\Phi_\nu = B_n,\ n\in\mathbb{N}.$$

Obviously (6.17) holds for n=1. Assume that (6.17) holds for $1\leq n\leq m$, hence $\bigcup_1^n\Phi_\nu = \sum_1^n B_\nu$ for $1\leq n\leq m$. Now we get successively the following series of equivalences:

(i) $s\in\Phi_{m+1} - \bigcup_1^m\Phi_\nu$,

(ii) $p(i,f(i),s)>0$ for some $i\in\Phi_m$ and $p(j,f(j),s)=0\ \forall j\in\bigcup_1^{m-1}\Phi_\nu$,

(iii) $p(i,f(i),s)>0$ for some $i\in\Phi_m - \bigcup_1^{m-1}\Phi_\nu = B_m$ and $p(j,f(j),s)=0$ $\forall j\in\bigcup_1^{m-1}\Phi_\nu$

(iv) $p(i,g_m(i),s)>0$ for some $i\in B_m$ and $s\notin\bigcup_1^m\Phi_\nu=\sum_1^m B_\nu$,

(v) $s\in B_{m+1}$.

Hence (6.17) holds for $n=m+1$. ⌋

Theorem 6.11 (cf.theorem 6.5). Let us assume

α) A is a compact metric space;

β) the sets $D(s), s\in S$, are closed;

γ) $p(s,\cdot,j)$ is l.s.c.;

δ) $r(s,\cdot)$ is u.s.c.

Then there exists in the discounted case a $\overline{p}$-optimal stationary plan.

Remark. One may generalize the stationary model to an instationary Markovian model with time-dependent discount rates by assuming that r_n is of the form $r_n=\beta_n r_{n-1}$, $n\geq 2$, and $0\leq\beta_n\leq 1$. Of course, this model is but a special case of our general Markovian model. The cases (EN),(EP) and (C) may be characterized as in the stationary model. E.g. case (C), which has been studied by Furukawa (67), is equivalent to "r_1 bounded and $\sum_n\prod_1^n\beta_\nu<\infty$".

7. Models with incomplete information

In the literature (c.f. e.g. Dynkin (65), Aoki (65) and (67), Sirjaev (67)) there have been studied so-called *models with incomplete information* (abbreviated by MII). These are decision models $DM'=(S',A,D',(p_n'),(r_n'))$ +) of the following kind:

(i) The countable state space S' is the product of the set S of the so-called *observable states* and the set C of the so-called *concealed states*. $z:=(c_1,c_2,\ldots,c_n)$ will denote the *concealed history* at time n, while $h:=(s_1,a_1,\ldots,s_n)$ will denote the *observable history*.

(ii) The action space is assumed to be countable. (In chapter II arbitrary sets S,C,A are admitted.)

(iii) The set $D_n'(h,z)$ of actions available at time n when the history (h,z) occurs, does not depend on the concealed history z, i.e.

$$(7.1) \qquad D_n'(h,z) = D_n(h),\ n\in\mathbb{N},\ (h,z)\in H_n'.$$

As a consequence we have $H_n'=H_n\times C^n$, where H_n is the set of observable histories.

(iv) Only those plans $f'\in\Delta'$ may be used, for which $f_n:(S\times C)^n\to A$ does not depend on z. Let Δ_S' be the (non-empty) set of all such plans.

(v) The notion of $\overline{p}$-optimality is not based on

$G:=\sup\limits_{f'\in\Delta'} G_{f'}'$, but on

$$(7.2) \qquad V':=\sup_{f'\in\Delta_S'} G_{f'}'.$$

The idea behind a MII is obvious: the observer of the process can base his decisions at time n only on the observable history h. Dynkin (65) describes several applications of such models.

Let us call a plan $f'\in\Delta_S'$ *$\overline{p}_S$-optimal*, if $G_{f'}'=V'$.

Because of condition (v) we optimize G_f' only within the class of plans in Δ_S', and the model does not fit directly into our theory. On the other hand, MII's seem to be a substantial generalization of our model introduced in section 2, since the latter is obtained from the former by taking for C

+) All notions that refer to the model DM' will be distinguished from those referring to the model DM by a (').

a set consisting of just one point. However, it has been remarked by Hinderer (67),p.28, that the problem of optimization in the MII $(S',A,D',(p_n'),(r_n'))$ may be solved in a natural manner by a DM $(S,A,D,(p_n),(r_n))$ of the following structure:

α) $S,A,D_n(h),H_n$ are the elements already defined in the MII. As a consequence, the map $f'\to f$ from Δ_S' into Δ, defined by

(7.3) $$f_n(y):=f_n'(y,z),n\in\mathbb{N},y\in S^n,$$

is bijective. In the sequel, we shall not distinguish between f and f'.

β) Define the functions $q_n(h;z)$ recursively by

(7.4) $$q_1(s;c):=p(s,c)/\sum_{\gamma\in C}p(s,\gamma),\quad (s,c)\in S\times C,$$

and

(7.5) $$q_{n+1}(h,a,s;z,c):=q_n(h,z)\cdot p_n'(h,z,a,s,c)/\sum_{\zeta,\gamma}q_n(h,\zeta)p_n'(h,\zeta,a,s,\gamma)$$
$$n\in\mathbb{N},\ (h,a,s;z,c)\in H_{n+1}'.$$

Obviously $q_n(h;z)$ will be interpreted as the conditional probability for the concealed history z, given the observable history h.

Now we can define the transition law and the reward function in the model DM by

(7.6) $$p(s):=\sum_c p'(s,c),$$
$$p_n(h,a,s):=\sum_{z,c}q_n(h;z)p_n'(h,z,a,s,c),n\in\mathbb{N},(h,a,s)\in H_{n+1},$$

(7.7) $$r_n(h,a):=\sum_z q_n(h;z)r_n'(h,z,a),n\in\mathbb{N},\ (h,a)\in K_n.$$

The probability measure on $\bigotimes_1^\infty\mathcal{P}(S')$, determined by $f\in\Delta'$ will be denoted by P_f'. Let η_n' and φ_n' denote the coordinate variables on $S'^{\mathbb{N}}$ for y_n and z_n, respectively. If we define

$$q_{nf}(y;z):=q_n(h_{nf}(y),z),$$

then we derive from the recursive definition of q_n, that

(7.8) $$q_{nf}(y;z):=P_f'(\varphi_n'=z|\eta_n'=y),n\in\mathbb{N},y\in S^n,z\in C^n.$$

Moreover, (7.5) and (7.6) imply

(7.9) $$q_{nf}(y;z)\cdot p_{nf}'(y,z,s,c)=q_{n+1,f}(y,s;z,c)p_{nf}(y,s).$$

From (7.8) and (7.6) we get

(7.10) $$P_f'(\eta_n'=y)=P_f(\eta_n=y),\ n\in\mathbb{N},y\in S^n.$$

The construction of $\bar{p}_S$-optimal plans cannot be based on the OE for the sequence of functions G_n',
$G_n'(h,z) := \sup_{f\in\Delta_n'(h,z)} G_{nf}'(y,z)$, but on a modified OE for the sequence of functions

$$V_n'(h) := \sup_{f\in\Delta_{nS}'(h)} V_{nf}'(y),\ n\in\mathbb{N},\ h\in H_n,$$

where

$$\Delta_{nS}'(h) := \{f\in\Delta_S : f_\nu(y_\nu) = a_\nu,\ \ 1\le\nu<n\},$$

and

$$V_{nf}'(y) := E_f'[\sum_n^\infty r_{\nu f}'\circ(\eta_\nu',\varphi_\nu')\,|\,\eta_\nu' = y],\ n\in\mathbb{N}, f\in\Delta_S, y\in S^n.$$

While in the literature the OE for (V_n') is usually derived without reference to the OE for DM's we shall derive it by means of the model $(S,A,D,(p_n),(r_n))$ introduced above. For that purpose we shall prove

<u>Theorem 7.1.</u> $V'=G$, $V_n'=G_n$, $n\in\mathbb{N}$.

Proof. Since $\Delta_{nS}'(U)=\Delta_n(h)$ (except for the bijection (7.3)), it is sufficient to show $V_{nf}'=G_{nf}$ for $n\in\mathbb{N}, f\in\Delta$. ($V_f'=G_f$ follows then from proposition 3.5.) From theorem A4 in appendix 2, lemma 3.6, and the proof thereof we conclude that

$$V_{nf}'(y)=\sum_n^\infty E_f'[r_{\nu f}\circ(\eta_\nu',\varphi_\nu')\,|\,\eta_n'=y]=$$
$$=\sum_n^\infty \sum_z q_{nf}(y;z)E_f'[r_{\nu f}\circ(\eta_\nu',\varphi_\nu')\,|\,\eta_n=y,\varphi_n=z]=$$
$$=\lim_{k\to\infty}\sum_z\sum_{\nu=n}^{n+k} q_{nf}(y;z)E_f'[\ldots|\ldots]=\lim_k\sum_z q_{nf}(y;z)(\Lambda_{nf}'^{k}0)(y,z),$$

where

$$\Lambda_{nf}'^{k}0 := \Lambda_{nf}'\ldots\Lambda_{n+k,f}'0.$$

Therefore lemma 3.6 implies $G_{nf}=V_{nf}'$ if we can show that

$$(7.11)\quad \sum_{z\in C^n} q_{nf}(y;z)(\Lambda_{nf}'^{k}v)(y,z)=(\Lambda_{nf}^{k}v)(y),\ n\in\mathbb{N}, k\in\mathbb{N}_0, y\in S^n,$$

for any function $v:S^{n+k+1}\to\overline{\mathbb{R}}$ that is bounded from above or from below in the case (EN) or (EP), respectively. Assertion (7.11) is verified by induction on k. For $k=0$ and any $n\in\mathbb{N}$ we have $\sum_z q_{nf}(y;z)\Lambda_{nf}'v(y,z)=\sum_z q_{nf}(y;z)[r_{nf}'(y,z)+$

$$+\sum_{s,c} p_{nf}'(y,z,s,c)v(y,s)]=r_{nf}(y)+\sum_s p_{nf}(y,s)v(y,s)=\Lambda_{nf}v(y),$$

hence the assertion is true for $k=0$ and any $n\in\mathbb{N}$. Now we shall assume that (7.10) holds for $k=1,2,\ldots,m$ and any $n\in\mathbb{N}$. Then, using (7.9), we get

$$\sum_z q_{nf}(y;z)(\Lambda'^{m+1}_{nf}v)(y,z)=$$

$$=\sum_z q_{nf}(y;z)[r'_{nf}(y,z)+\sum_{s,c} p'_{nf}(y,z,s,c)(\Lambda'^{m}_{n+1,f}v)(y,z,s,c)]=$$

$$=r_{nf}(y)+\sum_s p_{nf}(y,s)\sum_{z,c} q_{n+1,f}(y,s;z,c)(\Lambda'^{m}_{n+1,f}v)(y,z,s,c) =$$

$$=r_{nf}(y)+\sum_s p_{nf}(y,s)(\Lambda^{m}_{n+1,f}v)(y,s)=(\Lambda^{m+1}_{n,f}v)(y).$$

Hence (7.11) is true for $k=m+1$ and any $n\in\mathbb{N}$. ⌋

After having established theorem 7.1 we can derive results for MII from the results in previous sections. Let us mention the following.

Theorem 7.2 (cf. Dynkin (65)) The sequence (V'_n) satisfies the (modified) OE

$$(7.12)\qquad V'_n(h)=\sup_{a\in D_n(h)}\sum_z q_n(h;z)[r'_n(h,z,a)+ +\sum_s V'_{n+1}(h,a,s)\sum_c p'_n(h,z,a,s,c)],\ n\in\mathbb{N}, h\in H_n.$$

From (7.10) and theorem 5.1 we get

Theorem 7.3. The plan $f\in\Delta'_S$ is $\overline{P}_S$-optimal iff $f_n(y)$ is a maximum point of

$$a\to\sum_z q_{nf}(y,z)[r'_n(h_{nf}(y),z,a)+\sum_s V'_{n+1}(h_{nf}(y),a,s)\sum_c p'_n(h_{nf}(y),z,a,s,c)]$$

whenever $P'_f(\eta'_n=y)>0$.

Remark. The model of Dynkin (65) is in so far a special case of our model DM', as there $D_n(h)$, $p'_n(h,z,a,\cdot)$ and $r'_n(h,z,a)$ do not depend on the actions $a_1,a_2,\ldots,a_{n-1}$. On the other hand, Dynkin does not make assumptions like (EN) and (EP), but he assumes, that for any $f\in\Delta'_S$ the following conditions are fulfilled (the model is then called 'regular'):

(i) $(\Lambda'^{k}_{nf}0)$ exists for any $n\in\mathbb{N}, k\in\mathbb{N}_o$,

(ii) $\overline{V}_{nf}:=\lim_k(\Lambda'^{k}_{nf}0)$ exists for any $n\in\mathbb{N}$,

(iii) $\overline{V}_{nf} = \Lambda'_{nf}\,\overline{V}_{n+1}$, $n\in\mathbb{N}$.

(As we have shown, these assumptions are satisfied in either of the cases (EN) and (EP).) Since under these assumptions $\sum r'_{nf}\circ(\eta'_n,\varphi'_n)$ does not necessarily exist, Dynkin defines V'_{nf} by $\overline{V}_{nf}$, which is consistent with our definition in cases (EN) and (EP). The main result of Dynkin (65) is a condition which guarantees that (V_n') satisfies the OE and the existence of a plan $f \in \Delta_S'$ for which

(7.13) $V_{nf}'(y) \geq V_n'(h_{nf}(y)) - \varepsilon$ for $n \in \mathbb{N}$ and *all* $y \in S^n$.

Furthermore it is shown that the last mentioned condition is satisfied in case (C). In Hinderer (67) plans with (essentially) property (7.13) were called *strongly* ε-*optimal*, and some results were derived that are generalizations of results of Blackwell (65) on optimal plans. However, as pointed out in remark 10 of section 2, the notion of $\bar{p}$-optimality seems to be more appropriate in applications than the notion of optimality or strongly optimality.

Now we shall give a more specific application of a MII. In many problems of optimal control the state of the system cannot be observed exactly because of the presence of *random noise* and also the distribution of the state at time n+1 does not only depend on previous states and actions but also on some other source of random noise. Many problems of this kind have been treated by Aoki (67). We shall show how a simple problem of this kind (cf. Aoki (67),p.21) may be treated as a model with incomplete information. Let us make the following assumptions.

(i) There is known the counting density ρ of the concealed state c_1. At time n, the actual but concealed state c_n and some random noise β_n produce the observable state

$$s_n := \Gamma_n(c_n, \beta_n) \qquad \text{('observation equation')}.$$

(ii) If the concealed state c_n occurred and if action a_n has been taken, then some random noise ε_n produces the new concealed state

$$c_{n+1} := F_n(c_n, a_n, \varepsilon_n) \qquad \text{('plant equation')}.$$

(iii) The family $(\beta_n, \varepsilon_n, n \in \mathbb{N})$ of noise random variables is independent, and β_n and ε_n have counting densities u_n and v_n, respectively.

(iv) The set $D_n'(h,z)$, $h:=(s_1,a_1,\ldots,s_n)$, $z:=(c_1,\ldots,c_n)$, does not depend on the concealed history z.

(v) Only those plans f are admitted for which $f_n(y,z)$ does not depend on the concealed history z.

In order to describe the problem by a MII $(S\times C, A, D', (p_n'), (r_n'))$, we only have to define

$p(s,c) := \rho(c)\cdot\sum_b u_1(b)$, where the sum is extended over those b for which $\Gamma_1(c,b)=s$;

$p_n(h,z,a;s,c) := \sum_e v_n(e)\cdot\sum_b u_{n+1}(b)$, where the sums are extended over those e and b for which

$F_n(c_n,a,e)=c$ and $\Gamma_{n+1}(c,b)=s$, respectively.

Remark. An analysis of Markovian models with incomplete information and general state and action space is given by Sawaragi and Yoshikawa (70). It should be clear, that a similar study is possible for non-stationary models.

8. Examples of special models

Example 1. Models with *finite horizon* $N \in \mathbb{N}$ are models where $r_n \equiv 0$ for all $n>N$.

Models in real life often have finite horizons. The theoretical basis for models with finite horizon simplifies in as much as the OE (3.15) reduces to the recursive relation

$$(8.1) \qquad G_n = U_n G_{n+1}, \; 1 \leq n \leq N,$$
$$G_n \equiv 0 \text{ for } n>N.$$

One should pay attention to the fact that a Markovian model with finite horizon N, in which all components (with the exception that $r_n=0$ for $n>N$) are independent of n is not a stationary model (except in trivial cases). Hence, if there exists a $\overline{p}$-optimal plan, then there exists (at least in case (C), cf.theorem 6.3) a $\overline{p}$-optimal Markov plan but in general there does not exist a $\overline{p}$-optimal stationary plan.

For N large, the computation time for a numerical solution becomes in general forbidding. Therefore one 'approximates' models with finite horizon by appropriate models with infinite horizon. This necessitates a careful analysis of the computational procedures for solving the (now non-recursive) OE and of the approximation error.

Example 2. Many models of *deterministic* dynamic programming result from our model if we spezialize (p_n) in such a way, that p is concentrated at some given point $x \in S$ and such that $p_n(h,a,\cdot)$ is concentrated at some point $\varphi_n(h,a)$, where the so-called transition operators $\varphi_n : K_n \to S$ are known. If δ_{ij} denotes the Kronecker symbol, then we have

$$(8.2) \qquad p(s) = \delta_{sx}, \qquad s \in S,$$
$$p_n(h,a,s) = \delta_{s,\varphi_n(h,a)}, \; n \in \mathbb{N}, (h,a,s) \in H_{n+1}.$$

Let f be any plan in Δ. Then the decision process determined by f degenerates into a single sequence $\omega_f \in \Omega$, i.e. there is some point ω_f such that $P_f(\{\omega_f\})=1$. The point ω_f is determined by $\omega_f := (s_{nf})$, where s_{nf} is defined recursively together with the sequence (a_{nf}) of actions occurring under f by

$$s_{1f} := x$$
$$a_{nf} := f_n(s_{1f}, s_{2f}, \ldots, s_{nf}), n \in \mathbb{N},$$
$$s_{n+1,f} := \varphi_n(s_{1f}, a_{1f}, s_{2f}, \ldots, a_{nf}), n \in \mathbb{N}.$$

By induction one easily verifies that

$$y_{nf} := (s_{1f}, s_{2f}, \dots, s_{nf})$$

depends only on $a_{1f}, a_{2f}, \dots, a_{n-1,f}$ (and on x). The (expected) total reward under f is

$$G_f = R_f(\omega_f) = \sum r_{nf}(y_{nf}, f_n(y_{nf})),$$

hence G_f depends on f only by means of the sequence $z_f := (a_{nf})$ of actions occurring under f, i.e. $G_f = \Gamma_1(z_f)$. For any sequence $z_n = (a_1, \dots, a_n) \in A^n$ of actions a_ν we define the corresponding history at time n by $h_{n+1}(z_n) := (x, a_1, \varphi_1(x, a_1), \dots, \varphi_n(x, a_1, \dots, a_n))$, provided that

(8.3) $$a_1 \in D_1' := D_1(x),$$

$$a_\nu \in D_\nu'(a_1, a_2, \dots, a_{\nu-1}) := D_\nu(h_\nu(z_{\nu-1})), 1 < \nu \leq n.$$

Let Z_n denote the set of points $(a_1, a_2, \dots, a_n)$ that satisfies (8.3), and let Δ' be the set of sequences (a_n) that satisfy (8.3) for all $n \in \mathbb{N}$.
Δ' may be called the set of admissible action sequences, if the process starts in x. If f is an admissible plan, then obviously $z_f \in \Delta'$. Moreover, it is not difficult to see that the map $f \to z_f$ from Δ to Δ' is surjective. Hence lemma 3.1 tells us that the maximal (expected) reward is

$$G = \sup_{f \in \Delta} G_f = \sup_f \Gamma_1(z_f) = \sup_{z \in \Delta'} \Gamma_1(z).$$

Here becomes apparent a difference between deterministic and stochastic optimization, important for applications. +) While the later requires to maximize some function on a set of sequences of *functions*, in the deterministic models one needs only to maximize some function on a set of sequences of elements. This difference is sometimes obscurred if one wants to solve the deterministic problem for *arbitrary* x, e.g. in order to construct an *optimal* plan f (in the sense of remark 1 of section 2) by 'composing' f from x-optimal sequences $(a_n(x))$.

+) The difference disappears in the abstract setting, since any function may be interpreted as a point in a function space.

The OE reads as follows:

$$(8.4)\qquad G_n(h) = \sup_{a\in D_n(h)} \left[r_n(h,a)+G_{n+1}(h,a,\varphi_n(h,a))\right],$$

$$n\in \mathbb{N}, h\in H_n.$$

For the application of previously given results one should note that the set T_{nf} consists of the only point y_{nf}. If the model is Markovian, then φ_n does not depend on $s_1,a_1,\ldots,a_{n-1}$, i.e. $\varphi_n(h,a)=\varphi_n(s_n,a)$ for $n\in\mathbb{N}$, $(h,a)\in K_n$. Therefore $a_{nf}=f_n(s_{nf})$ and $s_{n+1,f}=\varphi_n(s_{nf},a_{nf})$ for any Markov plan f. Hence T'_{nf} consists of the only point s_{nf}. One should note that also in stationary models and for stationary plans f the set T'_{nf} does depend on n. Moreoever, s_{nf},a_{nf},T_{nf} depend on the choosen initial state x.

We mention the following result in the statement of which we shall use the term 'x-optimal' instead of '$\overline{p}$-optimal', if p is concentrated in $x\in S$.

Theorem 8.1 (cf.Theorem 6.4). Let the model be deterministic and Markovian and assume case (C). Then for any Markov plan f the following statements are equivalent:

α) f is x-optimal;

β) $f_n(s_{nf})$ is a maximum point of the map $a\to r_n(s_{nf},a)+G_{n+1}(\varphi_n(s_{nf},a)), n\in\mathbb{N}$;

γ) $G_n(s_{nf})=r_n(s_{nf},a_{nf})+G_{n+1}(s_{n+1,f}), n\in\mathbb{N}$.

Example 3. Here we mention the well-known fact (cf.e.g. White (69),p.32) that the classical problem of *maximizing a real function of a finite number of variables under constraints* may be formulated by a deterministic and in generel non-Markovian DM. (Whether the methods of dynamic programming can successfully be applied to such problems depends strongly on the structure of the function and of the constraints.)

Let A be an arbitrary non-empty set, $V:A^k\to\mathbb{R}$ be the function to be maximized. The constraints are usually written in the form

$$(8.5)\qquad v_i(a_1,a_2,\ldots,a_k)\leq 0,\quad 1<i\leq m.$$

Let us write them in the vector form

$$(8.6)\qquad v(a_1,a_2,\ldots,a_k)\leq 0.$$

(This is already the most general form of constraints; for any set of compatible constraints is determined by the set B of points $a \in A^k$ which satisfy all constraints, and B is determined by the condition $1-1_B(a_1,a_2,\dots,a_k) \leq 0$.)

It is not difficult to see that the problem may be described by the deterministic DM $(S,A,D,x,(\varphi_n),(r_n))$ (cf.example 2) where

(i) $S = A$

(ii) x is an arbitrary (and completely irrelevant)point in S,

(iii) $D_n(x,a_1,\dots,s_n):=\Gamma_n(a_1,a_2,\dots,a_{n-1})$ where

$$\Gamma_1:=pr(B\to A)=\{a_1 \in A: v(a_1,c)\leq 0 \text{ for some } c\in A^{k-1}\},$$

$$\Gamma_n(a_1,\dots,a_{n-1}):=(a_1,\dots,a_{n-1})\text{-section of } pr(B\to A^n)=$$

$$=\{a_n \in A: v(a_1,a_2,\dots,a_n,c)\leq 0 \text{ for some } c\in A^{k-n}\}, \quad 1<n<k,$$

$$\Gamma_k(a_1,\dots,a_{k-1}):=(a_1,\dots,a_{k-1})\text{-section of } B=$$

$$=\{a_k \in A: v(a_1,\dots,a_{k-1},a_k)\leq 0\},$$

$$\Gamma_n := A \ , \ n>k.$$

(iv) $\varphi_n(a_1,s_1,\dots,s_n,a_n)=a_n$.

(v) $r_n \equiv 0$ for $n \neq k$,

$$r_k(x,a_1,s_2,\dots,s_k,a_k)=V(a_1,a_2,\dots,a_k).$$

A plan f is admissible iff $(a_{1f},a_{2f},\dots,a_{nf})\in B$. Furthermore $G_f=V(a_{1f},a_{2f},\dots,a_{kf})$, hence our problem is described by the DM defined above.

The OE is here (as in all deterministic DM with finite horizon) an easy consequence of lemma 3.2 and yields for

$$G_n(h) = G_n(a_1,a_2,\dots,a_{n-1}), \ G_1(x) = G_f$$

the recursion

$$(8.7) \quad \begin{cases} G_n(a_1,a_2,\dots,a_{n-1})=\sup\limits_{a\in\Gamma_n(a_1,\dots,a_{n-1})} G_{n+1}(a_1,a_2,\dots,a_n) \\ \qquad\qquad 1\leq n<k, \\ G_k = V. \end{cases}$$

<u>Example 4</u>. In applications, e.g. in control theory, there are considered models $(S,A,D,(p_n),(r_n))$ together with the instruction to stop the process at the earliest time n for which h_n belongs to some given (possibly empty) set $\Phi_n \subset H_n$. We shall speak of a DM with the *stopping-sets* Φ_n. (One could allow Φ_n to depend on (h_{n-1},a_n); but if we define $\overline{\Phi}_n:=\{h_n\in H_n: h_n\in\Phi_n(h_{n-1},a_n)\}$, then we are back in our original case.) We shall show that one can describe this problem by

changing the reward functions r_n to functions r'_n in the following manner. Let us consider the set B_n of histories $h \in H_n$ which do not result in a stop, i.e.

$$(8.8) \qquad B_n := \{h_n \in H_n : h_\nu \notin \Phi_\nu \text{ for } 1 \leq \nu \leq n\}.$$

Then we just have to define

$$(8.9) \qquad r'_n(h,a) := r_n(h,a) \cdot 1_{B_n}(h), n \in \mathbb{N}, (h,a) \in K_n.$$

If $h \notin B_n$, then $(h,a_n,\ldots,s_m) \notin B_m$ for all $m>n$ and for all $(h,a_m,\ldots,s_m) \in H_m$, hence $r_m(h,a_n,\ldots,s_m,a_m)=0$, hence

$$(8.10) \qquad G'_n(h) = 0 \ , \ n \in \mathbb{N}, \ h \notin B_n.$$

The OE reads now

$$(8.11) \qquad G'_n(h) = \sup_{a \in D_n(h)} \left[r_n(h,a) + \sum_j p_n(h,a,j) G'_{n+1}(h,a,j)\right] \cdot 1_{B_n}(h), \quad n \in \mathbb{N}, \ h \in H_n.$$

As a practical problem we mention the task to proceed to a given point $c \in S$ as rapidly as possible (cf. Boudarel, Delmas, Guichet (68), p.23). Here $\Phi_n : K_{n-1} \times \{x\}$, $r_n := -1$, $B_n := \{h \in H_n : s_\nu \neq x \text{ for } 1 \leq \nu \leq n\}$, and the minimal expected time $Z_n(h) := -G'_n(h)$ from time n and history h until stopping satisfies the OE

$$(8.12) \qquad Z_n(h) = \inf_{a \in D_n(h)} \left[1 + \sum_j p_n(h,a,s) Z_{n+1}(h,a,s)\right] \cdot 1_{B_n}(h), \quad n \in \mathbb{N}, h \in H_n.$$

Note that (8.12) is valid whether the process stops with probability one or not.

Example 5. Another kind of a stopping problem arises when together with a model $(S,A,D,(p_n),(r_n))$ there is given the instruction to perform before time n+1 a trial which with probability $\alpha_n(h,a)$ tells us to stop the process (if it has not stopped before). We shall speak of a *model with random termination*. Note that example 4 does not fit into example 5, though we can compute in example 4 the probability $\alpha_n(h,a) := \sum_{j \in \Phi_n} p_n(h,a,j)$ that the process will stop at time n+1, if (h,a) has occurred and if we have not stopped before. The difference lies in the fact that in example 4 the decision to stop at time n+1 depends on the state s_{n+1} of the unstopped process, while in example 5 it depends on a random device that is independent of s_{n+1}.

We can describe the problem by a new model $(S',A,D',(p_n'),(r_n'))$ defined as follows: At first we adjoin to S some point $x \notin S$, which we call the *stopping state*, hence $S':=S+\{x\}$; further

$$D_1'(s):=\begin{cases} D_1(s) & , s \neq x \\ A & , s=x \end{cases} \quad \text{and}$$

$$D_n'(h):=\begin{cases} D_n(h) & , h \in H_n \\ A & , h \in H_n'-H_n \end{cases} \quad ;$$

$$p_n'(h,a,s)=\begin{cases} p_n(h,a,s)(1-\alpha_n(h,a)) & , s \neq x \\ \alpha_n(h,a) & , s=x \end{cases} \quad ;$$

$$r_n'(h,a):=r_n(h,a)\cdot 1_{H_n}(h).$$

The OE equation reads now

(8.13) $G_n'(h)=$

$$=\sup_{a \in D_n(h)} \left[r_n(h,a)+(1-\alpha_n(h,a))\sum_{j \in S} p_n(h,a,s)G_{n+1}'(h,a,s)\right] 1_{H_n}(h),$$
$$n \in \mathbb{N}, h \in H_n'.$$

The special case where the original model is Markovian has been considered by Boudarel, Delmas, Guichet (68),p.66. If the original DM is stationary with discount rate β, and if $\alpha_n(h,a)=\alpha$, then the model DM' is equivalent to a stationary model with discount rate $\alpha\beta$. This case has been considered by Howard (60/65), p.74.

Example 6. There are economical situations which are described by the behavior of two different systems: the so-called *exterior system* moves through the state space S according to some probability law, uncontrollable by the observer, while the so-called *interior* system moves through the action space, completely controllable by the observer. As a practical example we mention a production - and inventory problem analysed by Kall (64) and generalized by Hinderer (67). There the exterior system is the economical environment and the interior system is the totality of inventories and production capacities.

Formally, a situation with exterior and interior system is described by a DM $(S,A,D,(p_n),(r_n))$, where $p_n(h,a_n,\cdot)$ does not depend on $a_1,a_2,\dots,a_n$. Hence, the probability measures P_f that describe the decision process determined by f, do not depend on f, i.e. $P_f=P$, and the exterior process is the process of states (ζ_n) on $(\Omega,\mathfrak{A},P)$, whereas the interior process, generated by the plan f, is the process of actions $(f_n \circ (\zeta_1,\zeta_2,\dots,\zeta_n))$ on $(\Omega,\mathfrak{A},P)$. Since in general the reward function r_n are *not* independent of $s_1,s_2,\dots,s_n$, the decisions a_n to be taken will actually depend on the behavior of the exterior system, and G_f will actually depend on f.

Example 7. It has been known for some time (cf.e.g. Blackwell (62a), Matula (64), Black (65)) that problems of *optimal search* may be handled by dynamic programming methods. We shall now formulate some of these problems by decision models.

Problem 1. (cf. Matula (64), Black (65)).
A single target is in one of countably many cells and cannot move. The *search problem* is described by the tupel $(I,(q_i),(\gamma_i),(c_i))$ of the following meaning:

(i) I is the countable set of cells, the so-called *target space*;

(ii) q_i denotes the probability that the target is in cell i;

(iii) γ_i is the probability that a target, being in cell i, will not be detected on a single inspection there;

(iv) $c_i \geq 0$ is the cost of one inspection in cell i.

The problem is described by the following DM $(S,A,D,(p_n),(r_n))$ whereby we allow for theoretical reasons that the inspection process goes on forever.

$S:=\{0,1\}$, where $s_n=0$ or $s_n=1$ denotes an unsuccessful or successful inspection, at time n, respectively;

$A=I$, where a_n denotes the cell inspected at time n;

$D_n(h)=A \;\forall n,h$;

$p(0):=1$ and $p_n(h,a,1):=q_a(1-\gamma_a)$;

$r_n(h,a):=-c_a \,\delta_{o,\sum_1^n s_\nu}$.

Because of the form of r_n the model is *not* Markovian. Furthermore we are in case (EN). Moreover, example 7 is a particular case of example 4.

Problem 2. (cf. Beck (64),(65)). We are looking for a target that is situated on the real line according to a probability measure Q. (For simplicity we assume $Q(\{0\})=0$.) We start at the origin in the positive direction, and if the target is not found before reaching a_1, we reverse the direction and explore the other half of the real line up to some point $a_2 \leq 0$, unless we have found the target, etc. Hence any sequence $(a_1, a_2, \ldots)$ satisfies $0 \leq a_1 \leq a_3 \leq a_5 \leq \ldots$ and $0 \geq a_2 \geq a_4 \geq \ldots$. One is looking for a plan that minimizes the expected distance travelled.

Here we may use the DM

$S := \{0,1\}$, where $s_n = 0$ denotes an unsuccessful n-th 'journey';

$A := \mathbb{R}$, where a_n denotes the turning point of the n-th journey;

$$D_n(h) = \begin{cases} (a_{n-2}, \infty), & \text{if n is odd,} \\ (-\infty, a_{n-2}), & \text{if n is even;} \end{cases} \quad a_{-1} := a_o := 0;$$

$$p(1)=0,\ p_n(h, a_n, 1) = \begin{cases} Q((a_{n-2}, a_n\rangle), & \text{if n is odd,} \\ Q(\langle a_n, a_{n-2})), & \text{if n is even.} \end{cases}$$

$$r_n(h, a_n) = -|a_n - a_{n-1}| \cdot \delta_{o, \sum_1^n s_\nu} .$$

Here again we have a non-Markovian problem.

Example 8. Problems of *optimal stopping* have attracted widespread attention during recent years. Breiman (64) defines such a problem by a tupel $(I, I_c, I_s, p, (p_{ij}), F, f)$ of the following meaning: The player starts in some point of the (countable) set I according to the initial distribution p; he moves through the states of I according to the Markov chain generated by the stochastic matrix (p_{ij}); whenever he is in state $i \in I$ he has at most the following two possibilities: either he may stop and receive the income $F(i) \in \mathbb{R}$ or he may continue the play by paying the entrance-fee $f(i) \in \mathbb{R}$; whenever the player arrives at some point $i \in I_s \subsetneqq I$, he *must* stop whereas he *must* continue whenever he arrives at some point $i \in I_c \subsetneqq I$, $I_c \cap I_s = \emptyset$.

We may formulate the described stopping problem as a non-stationary DM in the following way: $S := I$; $A := \{0,1\}$, where a=1 means "stop" whereas a=0 means "continue";

$$D_n(h) := \begin{cases} \{0,1\} & \\ \{0\} & \text{if } s_n \in \\ \{1\} & \end{cases} \begin{cases} I - I_s - I_c \\ I_c \\ I_s \end{cases} ; \quad p_n(h,a,j) := p_{s_n j} ,$$

$$r_n(h,a) := \prod_{\nu=1}^{n-1} \delta_{oa_\nu} \cdot [F(s_n)\delta_{1a} - f(s_n)\delta_{oa}].$$

A remarkable feature of this model is, that the pay-off structure but not the stochastic development of the system is affected by the actions. It is possible to show (cf. Breiman (64)) that the OE, whenever it holds, takes on the form

$$G_1(i) := \begin{cases} F(i) \\ -f(i) + \sum_j p_{ij} \cdot G_1(j) \\ \max[F(i), -f(i) + \sum_j p_{ij} G_1(j)] \end{cases} \quad \text{if } i \in \begin{cases} I_s \\ I_c \\ I - I_s - I_c \end{cases}$$

9. Randomized plans.

So far we have neglected the possibility that randomized plans ('mixed strategies') may yield a larger expected total reward than G. In this section we give a partial answer to this problem along the lines of Blackwell (65), Strauch (66), Hinderer (67). Since we want to defer measure-theoretic considerations to chapter II, we shall make for this section the *general assumption* that the action space A is countable.

Definition. A randomized plan π for the model $(S,A,D,(p_n),(r_n))$ is a sequence $\pi=(\pi_n)$, where $\pi_n(h,\cdot)$, $n\in\mathbb{N}, h\in\overline{H}_n:=S\times A\times\ldots\times S$, is a counting density on A such that

$$(9.1) \qquad \sum_{a\in D_n(h)} \pi_n(h,a) = 1,\ n\in\mathbb{N},\ h\in H_n.$$

The interpretation of a randomized plan π ist the following: If we decide to use π, and if under $\pi_1,\pi_2,\ldots,\pi_{n-1}$ at time n the history $h\in H_n$ occurs, then we select an action $a_n\in D_n(h)$ according to the counting-density $\pi_n(h,\cdot)$. Note that in contrast to deterministic plans the function π_n is allowed to depend on the whole history h, not only on the state history $y=(s_1,s_2,\ldots,s_n)$. A reduction similar to that mentioned in remark 8 of section 2 seems in general to be impossible.

If the counting density $\pi_n(h,\cdot)$ gives probability 1 to the point $g_n(h), n\in\mathbb{N}, h\in H_n$, then the randomized plan π is in fact a deterministic plan g, where g_n depends not only on the state-history but also on the action-history (cf. remark 8 of section 2). Let Δ' be the set of all plans g obtained in this way and let Δ^r be the *set of all randomized plans*. From (9.1) we infer that a sequence (g_n), $g_n:\overline{H}_n\to A$ belongs to Δ' iff

$$g_n(h)\in D_n(h), n\in\mathbb{N}, h\in H_n.$$

In the sequel we shall not distinguish between a plan $g\in\Delta'$ and the corresponding plan $\pi\in\Delta^r$. Hence we may regard Δ' as a subset of Δ^r. As already stated in remark 8 of section 2, the map $g\to f$, defined by

$$(9.2) \qquad f_n(y):= g_n(h_{nf}(y)),\ n\in\mathbb{N}, y\in S^n,$$

is a surjective map from Δ' to Δ.

If f is a deterministic plan, then the action at time n is described by the random variable $f_n \circ \eta_n$ on $(\Omega, \mathfrak{A}, P_f)$. If we use a randomized plan π, we have to use another probability space for the description of the decision process determined by π. We shall take as *sample space* the set $\overline{H} := S \times A \times S \ldots$ and as σ-algebra $\overline{\mathfrak{H}}$ the product-σ-algebra determined by the factors $\mathfrak{P}(S)$ and $\mathfrak{P}(A)$. For the description of the process we shall use the *coordinate variables*

$$\zeta_n(h) := s_n, \; n \in \mathbb{N}, h = (s_1, a_1, s_2, \ldots) \in \overline{H}, \tag{9.3}$$
$$\alpha_n(h) := a_n, \; n \in \mathbb{N}, \; h \in \overline{H}.$$

The history at time n is described by the random variable

$$\chi_n(h) := h_n, \; n \in \mathbb{N}, \; h \in \overline{H}. \tag{9.4}$$

For any $\pi \in \Delta^r$ there exists a unique probability measure P_π on $\overline{\mathfrak{H}}$ such that

$$P_\pi(\chi_n = h) = p(s_1) \prod_{\nu=1}^{n} [\pi_\nu(h_\nu, a_\nu) \cdot p_\nu(h_\nu, a_\nu, s_{\nu+1})] \quad n \in \mathbb{N}, h \in H_n. \tag{9.5}$$

P_π may also be described as the unique probability measure on $\overline{\mathfrak{H}}$ for which

(i) $P_\pi(\zeta_1 = s) = p(s) \;\; \forall s \in S$,

(ii) $P_\pi(\zeta_{n+1} = s \mid (\chi_n, \alpha_n) = (h, a)) = p_n(h, a, s)$, whenever $P_\pi((\chi_n, \alpha_n) = (h, a)) > 0$, and

(iii) $P_\pi(\alpha_n = a \mid \chi_n = h) = \pi_n(h, a)$, whenever $P_\pi(\chi_n = h) > 0$ holds.

<u>Lemma 9.1</u> Let π be any plan in Δ^r.

a) The set $H := \{h \in \overline{H} : h_\nu \in H_\nu \;\forall \nu \in \mathbb{N}\}$+) belongs to $\overline{\mathfrak{H}}$.

b) For any $n \in \mathbb{N}$ and $h \in H_n$ the conditional distribution of $(\alpha_n, \zeta_{n+1}, \alpha_{n+1}, \ldots)$ under the condition $[\chi_n = h]$ is concentrated on the h-section H_h of H, i.e.

$$P_\pi((\alpha_n, \zeta_{n+1}, \ldots) \in H_h \mid \chi_n = h) = 1. \tag{9.6}$$

Furthermore $P_\pi(H) = 1$.

Proof. a) Since $\mathfrak{A}_n := \mathfrak{P}(S) \otimes \mathfrak{P}(A) \otimes \ldots \otimes \mathfrak{P}(S)$ (2n-1 factors) equals $\mathfrak{P}(\overline{H}_n)$, we have $H_n \in \mathfrak{A}_n$ for any n, hence $H_n \times A \times S \times \ldots =: \hat{H}_n \in \overline{\mathfrak{H}}$, hence $\hat{H} = \bigcap_n \hat{H}_n \in \overline{\mathfrak{H}}$.

+) H is called the *set of admissible paths.*

b) We have $H_h = (\bigcap_{\nu>n} \hat{H}_\nu)_h = \bigcap_{\nu>n} (\hat{H}_\nu)_h$ and for $\nu \geq n$

$$
\begin{aligned}
&P_\pi((\alpha_n, \zeta_{n+1}, \ldots) \in (\hat{H}_{\nu+1})_h \mid \chi_n = h) = \\
&= P_\pi((\alpha_n, \zeta_{n+1}, \ldots, \zeta_{\nu+1}) \in (H_{\nu+1})_h \mid \chi_n = h) = \\
(9.7) \quad &= \sum_{(a_n, \ldots, s_{\nu+1})} \pi_n(h, a_n) p_n(h, a_n, s_{n+1}) \ldots \\
&\ldots p_\nu(h_\nu, a_\nu, s_{\nu+1}) \cdot 1_{H_{\nu+1}}(h, a_n, \ldots, s_{\nu+1}).
\end{aligned}
$$

Since $1_{H_{\nu+1}}(h, a_n, \ldots, s_{\nu+1}) = 1_{H_\nu}(h, \ldots, s_\nu) \cdot 1_{D_\nu(h, \ldots, s_\nu)}(a_\nu)$ and $\sum_{a_\nu} \pi_\nu(h, \ldots, s_\nu) \cdot 1_{D_\nu(h, \ldots, s_\nu)}(a_\nu) = 1$, the term (9.7) reduces to $\sum_{(a_n, \ldots, s_\nu)} \pi_n(h, a_n) \ldots p_{\nu-1}(h_{\nu-1}, a_{\nu-1}, s_\nu) \cdot 1_{H_\nu}(h, \ldots, s_\nu)$. Downward induction reduces this term to 1. ⌋

We remind of our convention to put r_n equal to zero on $\overline{H}_n \times A - K_n$. Hence under either of the assumptions (EN) and (EP) there exists the *total reward*

$$
\begin{aligned}
(9.8) \qquad R(h) :&= \sum_n r_n(h_n, a_n), \\
&= \sum_n r_n \circ (\chi_n, \alpha_n)(h), \quad h \in \overline{H},
\end{aligned}
$$

as an extended real valued random variable on $(\overline{H}, \overline{\mathfrak{H}}, P_\pi)$. Furthermore $R^+ \leq \sum_n \| r_n^+ \| < \infty$ in case (EN) and $R^- \leq \sum_n \| r_n^- \| < \infty$ in case (EP). Hence there exists the *expected total reward*

$$
(9.9) \qquad G_\pi := E_\pi R
$$

and for any $n \in \mathbb{N}$ and $h \in H_n$ the *conditional* expectation $V_{n\pi}(h)$ of the *reward* for the time period (n, ∞) under the condition that the history h occurred at time n:

$$
(9.10) \qquad V_{n\pi}(h) := E_\pi \left[\sum_{i=n}^{\infty} r_i \circ (\chi_i, \alpha_i) \mid \chi_n = h \right], \quad n \in \mathbb{N}, \pi \in \Delta^r, h \in H_n.
$$

Let g be a plan in Δ' and $f \in \Delta$ the plan determined by g by means of (9.2). One easily proves with the aid of lemma 3.6 and its analogue for randomized plans, that

$$
(9.11) \qquad G_{nf}(y) = V_{ng}(h_{nf}(y)), \quad n \in \mathbb{N}, \; y \in S^n,
$$

and hence also $G_f = G_g$.

There arise two questions:

(i) Do we have $\sup_{\pi\in\Delta^r} G_\pi > G$ or $\sup_{\pi\in\Delta^r} G_\pi = G$?

(ii) Does there even exist to any plan $\pi\in\Delta^r$ a deterministic plan $f\in\Delta$ such that $G_f \geq G_\pi$?

At first one verifies that the proofs of theorems 3.9, 3.10 and 3.11 carry over to the case of randomized plans, if we replace G_n by

(9.12) $$V_n := \sup_{\pi\in\Delta^r} V_{n\pi}, \quad n\in\mathbb{N},$$

the *maximal expected reward* during the time interval (n,∞), when randomized plans are used. Hence we have

Theorem 9.2. (i) The sequence (V_n) satisfies the OE

(9.13) $$V_n = \sup_{a\in D_n(\cdot)} \left[r_n(\cdot,a)+\sum_j p_n(\cdot,a,j)V_{n+1}(\cdot,a,j)\right], \quad n\in\mathbb{N}.$$

(ii) $$V := \sup_{\pi\in\Delta^r} G_\pi = \sum_s p(s)V_1(s).$$

(iii) In case (EP) the sequence (V_n) is the (termwise) smallest of those solutions (v_n) of the OE that satisfy

$$\|v_n\| \geq -\sum_n^\infty \|r_\nu^-\|, \quad n\in\mathbb{N}.$$

The comparison with theorem 3.11 shows that $V_n = G_n, n\in\mathbb{N}$, in case (EP). In case (EN) we get, using an idea of Strauch (66), a stronger result in theorem 9.4 below.

Lemma 9.3. Let π and σ be plans in Δ^r, and denote by $(\pi^n\sigma)$ the plan $(\pi_1,\pi_2,\ldots,\pi_n,\sigma_{n+1},\sigma_{n+2},\ldots)$. If $k\in\mathbb{N}$ and

$V_{k\sigma} \leq \overline{\lim}_n V_{k(\pi^n\sigma)}$, then $V_{k\sigma} \leq V_{k\pi}$ in case (EN).

Proof. For any $n>k$ we get, since $\rho_\nu := r_\nu \circ (\chi_\nu,\alpha_\nu) - \|r_\nu^+\| \leq 0$,

$$V_{k(\pi^n\sigma)} - \sum_k^\infty \|r_\nu^+\| = E_{\pi^n\sigma}\left[\sum_k^\infty \rho_\nu \mid \chi_k=\cdot\right] \leq E_{\pi^n\sigma}\left[\sum_k^n \rho_\nu \mid \chi_k=\cdot\right] =$$

$$= E_\pi\left[\sum_k^n \rho_\nu \mid \chi_k=\cdot\right] \downarrow E_\pi\left[\sum_k^\infty \rho_\nu \mid \chi_k=\cdot\right] = V_{k\pi} - \sum_k^\infty \|r_\nu^+\|.$$

Therefore $\overline{\lim}_n V_{k(\pi^n\sigma)} \leq V_{k\pi}$, and the assertion follows. $\lrcorner$

Theorem 9.4 (cf. Strauch (66)). In case (EN) there exists to any randomized plan σ a deterministic plan f such that $G_f \geq G_\sigma$ and $G_{nf}(y) \geq V_{n\sigma}(h_{nf}(y))$, $n\in\mathbb{N}, y\in S^n$.

Proof. For any $\sigma\in\Delta^r$ and $n\geq k$ we put

$$v_n(h,a) := E_\sigma\Big[\sum_n^\infty r_\nu o(\chi_\nu,\alpha_\nu)\,|\,(\chi_n,\alpha_n)=(h,a)\Big],\ (h,a)\in\overline{H}_n\times A.$$

From $P_\sigma((\chi_n,\alpha_n)\in H_n)=1$ follows $v_n(h,a)=0$ for $(h,a)\notin K_n$.
In case (EN) we have $V_{n\sigma}(h)=\sum_{a\in D_n(h)}\sigma_n(h,a)v_n(h,a)<\infty, h\in H_n$.

Hence there exists an action $g_n(h)\in D_n(h)$ for which $v_n(h,g_n(h))\geq V_{n\sigma}(h)$. For $h\notin H_n$ we define $g_n(h)$ arbitrary. According to lemma 9.3 and the equation $G_\sigma=\sum_s p(s)V_{1\sigma}(s)$ (cf.theorem 9.2) it is sufficient to show

$$V_{k\sigma}\leq V_{k(g^n\sigma)}\ ,\quad n>k. \tag{9.14}$$

Now we have at first

$$V_{k\sigma}(h)\leq v_k(h,g_k(h)) = V_{k(g^k\sigma)}(h),\quad h\in H_k.$$

Furthermore we get for $n>k$

$$V_{k(g^n\sigma)}(h) = E_g\Big[\sum_k^n r_\nu o(\chi_\nu,\alpha_\nu)\,|\,\chi_k=h\Big]+W,$$

where
$$W := E_{g^n\sigma}\Big[\sum_{n+1}^\infty r_\nu o(\chi_\nu,\alpha_\nu)\,|\,\chi_k=h\Big]=$$
$$= \sum_z P_g((\alpha_k,\zeta_{k+1},\dots\zeta_{n+1})=z\,|\,\chi_k=h)\cdot$$
$$\cdot\sum_{a\in D_{n+1}(h,z)}\sigma_{n+1}(h,z,a)v_{n+1}(h,z,a)\leq$$
$$\leq\sum_z P_g((\alpha_k,\dots,\zeta_{n+1})=z\,|\,\chi_k=h)\,v_{n+1}(h,z,g_{n+1}(h,z))$$
$$=E_{g^{n+1}\sigma}\Big[\sum_{n+1}^\infty r_\nu o(\chi_\nu,\alpha_\nu)\,|\,\chi_k=h\Big].$$

Therefore $V_{k(g^n\sigma)}\leq V_{k(g^{n+1}\sigma)}$, and induction on n implies (9.14). Equation (9.11) completes the proof. ⌋

From the remark after theorem 9.2 and theorem 9.4 we get

<u>Corollary 9.5</u>. In either of the cases (EN) and (EP) we have

$$V_n = G_n,\ n\in\mathbb{N} \text{ and}$$
$$\sup_{\pi\in\Delta^r} G_\pi = G.$$

The proof of theorem 9.4 indicates, that its statement fails in general in case (EP). More explicitely we look at the following *example:* $A=D_n(h)=\mathbb{N}, p=\delta_x$ for some $x\in S, r_n\equiv 0$ for $n>1$, $r_1(x,\cdot)$ real, positive and unbounded. Then there exists a plan $\pi\in\Delta^r$ such that $G_\pi=\sum_{a\in\mathbb{N}}\pi_1(x,a)r_1(x,a)=\infty$, but $G_f=r_1(x,f_1(x))<\infty$ for any $f\in\Delta$.

It seems to be unknown in which subcase of case (EP) the statement of theorem 9.4 is true. According to corollary 9.5 the problem reduces to find conditions under which the existence of a randomized $\overline{p}$-optimal plan implies the existence of a deterministic $\overline{p}$-optimal plan.

More results on randomized plans may be found in chapter II.

10. Dynamic programming under uncertainty.

Adhering to the terminology that is customary in economics, we shall speak of a DM with *uncertainty*+) (abbreviated by DMU) if the transition law (p_n) of the model is not completely known to the observer (including the complete uncertainty case in which nothing is known about the transition law). Models with completely known transition law with which we have dealt so far, will be called models with *risk*. Models with uncertainty which are obviously very important for applications, have been studied by many authors. In most cases, the Bayesian point of view is taken, e.g. in Sworder (66), Martin (67), Aoki (67). Wessels (68) considers also other approaches of statistical decision theory, e.g. min-max risk.

We do not enter in a detailed discussion of the subject but we shall show how to reduce models with complete uncertainty under the Bayesian approach to models with risk studied in previous sections. This will supply an important example of a general, non-Markovian DM, since - as will be seen - even stationary Markovian models with uncertainty will lead to non-Markovian models with risk.

The general situation for DMU's is similar to that in *statistical decision theory*. In the latter we are given

(i) a measurable space $(X,\mathfrak{S})$, the so-called sample space;

(ii) a family $\mathfrak{Q}=(Q_\vartheta,\vartheta\in\Theta)$ of probability measures on $\mathfrak{S}$;

(iii) a set E, the so-called set of decisions;

(iv) a map $L:\Theta\times E\to \mathbb{R}$, the so-called loss function, which is usually assumed to be non-negative.

A (deterministic) decision function is then a map $\delta:X\to E$, and decision functions are compared on the basis of their so-called *risk function*

$$\rho_\delta(\vartheta) := E_\vartheta L(\vartheta,\delta) = \int L(\vartheta,\delta(x))Q_\vartheta(dx), \vartheta\in\Theta.$$

The aim is to find a decision function δ for which ρ_δ is in some sense a 'small' element of the (in general not totally ordered) family of all risk functions. The *Bayesian* approach consists in introducing in this family a total ordering in the following manner: We select a convenient σ-algebra $\mathfrak{T}$ in the parameter space Θ and a so-called *a priori probability*

+) or *adaptive* DM

measure μ on $\mathfrak{T}$ which usually describes our preference pattern for the possible values of the parameter $\vartheta \in \Theta$. Then we compare different decision functions on the basis of their *mean risk*

$$\overline{\rho}_\sigma := \int \rho_\delta d\mu .$$

Now let us turn to a DMU. Since the transition law is not completely known, it depends on some parameter ϑ variing in an appropriate parameter space Θ. Hence, for any plan f, the corresponding probability measure P_f^ϑ depends on ϑ.

Comparing a DMU and a problem in statistical decision theory, we note the following difference: The loss function L depends both on the unknown parameter and the decision e, whereas the total reward depends only on the plan f; on the other hand, the probability measures Q_ϑ do not depend on the decision e, whereas the probability measure P_f^ϑ depends also on f. However, this difference does not matter as soon as we decide to compare in the DMU different plans f solely on the basis of

$$(10.1) \qquad \rho_f(\vartheta) := -G_f^\vartheta := -\int R_f dP_f^\vartheta , \quad \vartheta \in \Theta ,$$

for then ρ_f plays exactly the same role as the risk function ρ_δ in a statistical decision problem. In particular, we may use the Bayesian approach described above.

From now on we shall restrict ourselves to *Markovian* DMU's where in addition A is countable and D_n and p_n ($n \in \mathbb{N}$) - but not necessarily r_n - are independent of n, i.e. $D_n = D$, $p_n = q$. Moreover we treat only the case of *complete uncertainty*, i.e. we have $\vartheta := q$. For the following discussion it is convenient to regard ϑ as a (countably dimensional) vector with components $\vartheta(i,a,j) = q(i,a,j)$, $(i,a,j) \in C := \{(s,b,t) \in S \times A \times S : b \in D(s)\}$. Therefore ϑ is a point in the euclidean space $\mathbb{R}^C$. From $q \geq 0$, $\sum_j q(i,a,j) = 1$ follows that we have to take as parameter space the set

$$(10.2) \qquad \Theta := \{\vartheta \in \mathbb{R}^C : \vartheta(i,a,j) \geq 0, \sum_s \vartheta(i,a,s) = 1, \forall (i,a,j) \in C\}.$$

As σ-algebra $\mathfrak{T}$ in Θ we shall take the trace of the σ-algebra $\mathcal{B}_C$ of Borel sets in $\mathbb{R}^C$ on Θ, i.e. $\mathfrak{T} := \Theta \cap \mathcal{B}_C$. Since for any $(i,a,j) \in C$ the maps $\vartheta \to \vartheta(i,a,j)$ are projections from $\mathbb{R}^C$ into $\mathbb{R}$, one concludes easily that Θ belongs to $\mathcal{B}_C$. Finally we assume that some a priori probability measure μ on $\mathfrak{T}$ is given.

We summarize the data of our problem in the following

Definition. A *Bayesian decision model* (BDM) is a tupel $(S,A,D,(p_n),(r_n),\mu)$ of the following meaning.

(i) $(S\ ,A,D,(p_n),(r_n))$ is a Markovian DM in which

α) A is countable,

β) D_n and p_n are independent of n, i.e. $D_n=D, p_n=\vartheta, n\in\mathbb{N}$.

(ii) μ is a probability measure on $\mathcal{T}$, the a priori probability for ϑ.

As already mentioned above, we want to compare different plans f by means of their *μ-average of the total expected reward*

$$(10.3) \qquad \overline{R}_f := \int G_f^{\vartheta}\mu(d\vartheta).$$

In order that the definition of $\overline{R}_f$ makes sense we have to show

Lemma 10.1. For any $f\in\Delta$ the map $\vartheta\to G_f^{\vartheta}$ from Θ to $\overline{\mathbb{R}}$ is measurable.

Proof. Since $G_f^{\vartheta}:=\int R_f(\omega)P_f^{\vartheta}(d\omega)$ and since $(\vartheta,\omega)\to R_f(\omega)$ is measurable, it is sufficient according to a well-known theorem in measure theory (cf. e.g. Neveu (64), p.74), to show that $\vartheta\to P_f^{\vartheta}(B)$ is measurable for any set of a subsystem of $\mathcal{F}$ that generates $\mathcal{F}$ and that is closed under intersections. Hence it suffices to consider only sets B of the form

$$B = \{(\omega_n)\in\Omega:\ \omega_r=s_r,\ 1\le r\le n\},\ n\in\mathbb{N},\ (s_1,s_2,\dots,s_n)\in S^n.$$

Since $P_f^{\vartheta}(B)=p(s_1)\prod_{\nu=1}^{n-1}\vartheta(s_\nu,f_\nu(y_\nu),s_{\nu+1})$ and since the product of measurable functions is measurable, we have to verify that each of the maps $\vartheta\to\vartheta(i,a,j)$ from Θ to $\mathbb{R}$ is measurable. This follows easily since the last mentioned map is the restriction from $\mathbb{R}^C$ to Θ of a projection. ⌋

Corollary 10.2. For any $f\in\Delta$ the map $(\vartheta,B)\to P_f^{\vartheta}(B)$ is a transition probability from $(\Theta,\mathcal{T})$ to $(\Omega,\mathcal{F})$.

For a rigorous description of the Bayesian approach we shall use the measurable space $(\Gamma,\mathcal{G})$, defined as the product of the parameter space $(\Theta,\mathcal{T})$ and of the state space $(\Omega,\mathcal{F})$ of the decision processes. Let $W_f:=\mu\otimes P_f^{\bullet}$ be the unique measure on $\mathcal{G}$ determined by the probability measure μ and the transition probability $(\vartheta,B)\to P_f^{\vartheta}(B)$. The probability space $(\Gamma,\mathcal{G},W_f)$ describes the combined experiment in which at first 'nature' selects ϑ according to the distribution μ and then 'nature' selects (after the choice of the plan f by the observer) some path $\omega=(s_1,s_2,\ldots)$ according to the distribution P_f^{ϑ}. Let us introduce on $(\Gamma,\mathcal{G})$ the random variables

(10.4) $$\tilde{p}(\vartheta,\omega):=\vartheta \quad , \; \tilde{\eta}(\vartheta,\omega):=\omega, \; \eta_n(\vartheta,\omega):=(\omega_1,\omega_2,\ldots,\omega_n),$$

i.e. in the combined experiment the selection of the parameter, of the state history and of the state history at time n are described by $\tilde{p},\tilde{\eta}$ and $\tilde{\eta}_n$, respectively. Under W_f the distribution of $\tilde{p}$ is simply μ, the distribution of $\tilde{\eta}$ is

(10.5) $$\overline{P}_f: = \int P_f^{\vartheta}\mu(d\vartheta),$$

and the distribution of $\tilde{\eta}_n$ is given by

(10.6) $$W_f(\tilde{\eta}_n=y) = \int P_f^{\vartheta}(\eta_n=y)\mu(d\vartheta), n\in \mathbb{N}, y\in S^n$$
$$= \overline{P}_f(\eta_n=y).$$

The distribution $\overline{P}_f$ may be regarded as the mixture of the probability measures $P_f^{\vartheta},\vartheta\in\Theta$, with respect to the a priori distribution μ.

The total expected reward $\overline{R}_f$ in our BDM is given by

(10.7) $$\overline{R}_f: = E_f\; R_f\circ\tilde{\eta}: = \int R_f\circ\tilde{\eta}\, dW_f,$$

which reduces to

(10.8) $$\overline{R}_f = \int R_f d\overline{P}_f.$$

Of course, a plan f^* will be called $\overline{p}$-optimal for the given BDM, if $\overline{R}_{f^*}=\sup_{f\in\Delta} \overline{R}_f$.

It is obvious that we can reduce our BDM to a DM under risk if we can show that each of the probability measures $\overline{P}_f$ admits a representation $\overline{P}_f=\bigotimes_0^\infty \overline{p}_{nf}$ for some transition law $(\overline{p}_n)$, independent of f. Intuitively, one will try to take for $\overline{p}_n(h_{n+1})$ the expectation of $\vartheta(s_n,a_n,s_{n+1})$ with respect to a conditional distribution of $\tilde{p}$ under the condition that $\tilde{\eta}_n=y_n$ and with respect to a probability measure W_f for some $f\in\Delta_n(h_n)$. This approach is possible since the pertinent conditional distribution is independent of $f\in\Delta_n(h)$. In fact, let $h\in H_n$ be fixed, and take some $f\in\Delta_n(h)$. Then

(10.9) $$P_f^{\vartheta}(\eta_n=y)=p(s_1)\prod_1^{n-1}\vartheta(s_\nu,f_\nu(y_\nu),s_{\nu+1})$$

has the same value for all $f\in\Delta_n(h)$. Hence

(10.10) $$W_f(\tilde{p}\in B\,|\,\eta_n=y)=\int_B P_f^{\vartheta}(\eta_n=y)\mu(d\vartheta)\Big/\int_\Theta P_f^{\vartheta}(\eta_n=y)\mu(d\vartheta)$$

has the same value $\tau_n(h,B)$ for all $f\in\Delta_n(h)$. We shall call $\tau_n(h,\cdot)$ the *a posteriori distribution* of $\tilde{p}$ under h. τ_n is a transition measure from H_n to Θ, such that $\tau_n(h,\cdot)$ is a probability measure, if $W_f(\tilde{\eta}_n=y)>0$, and the identically vanishing measure, otherwise.

Theorem 10.3. Define the sequence $(\overline{p}_n)$ by $\overline{p}_o:=p$, $\overline{p}_1(h_2):=\int\vartheta(h_2)\mu(d\vartheta)$ +), and for $n\geq 2$, $(h,a,s)\in H_{n+1}$

$$\overline{p}_n(h,a,s):=\begin{cases}\int\vartheta(s_n,a,s)\tau_n(h,d\vartheta), & \text{if } \tau_n(h,\cdot)\not\equiv 0,\\ \vartheta(s_n,a,s), & \text{otherwise.}\end{cases}$$

Then $\overline{P}_f=\bigotimes_o^\infty\overline{p}_{nf}$ for any $f\in\Delta$.

Proof. At first it is clear that (p_n) is a transition law, since $\overline{p}_n\geq 0$ and $\sum_s\overline{p}_n(h,a,s)$ equals $\int(\sum_s\vartheta(s_n,a,j))\tau_n(h,d\vartheta)=1$ (by Fubini's theorem) or $\sum_s\vartheta(s_n,a,s)=1$, according as $\tau_n(h,\cdot)\not\equiv 0$ or $\tau_n(h,\cdot)\equiv 0$. It remains to show that

(10.11) $$\overline{P}_f(\eta_n=y)=\prod_o^{n-1}\overline{p}_{\nu f}(y_{\nu+1}),\, n\in\mathbb{N}, y\in S^n.$$

From (10.6) follows immediately that (10.11) is true for n=1. Now let (10.11) be true for some $n\in\mathbb{N}$, and consider some $(y,s)\in S^{n+1}$. If $\overline{P}_f(\eta_n=y)=0$, then $\overline{P}_f(\eta_{n+1}=(y,s))=0$ and $\prod_o^n\overline{p}_{\nu f}(y_{\nu+1})=\overline{P}_f(\eta_n=y)\overline{p}_{\nu f}(y,s)=0$. Now we assume $\overline{P}_f(\eta_n=y)>0$, and put $h:=h_{nf}(y)$. Then $\tau_n(h,\cdot)\not\equiv 0$. The joint distribution of $\tilde{p}$ and $\tilde{\eta}_n$ under W_f is determined by $W_f(\tilde{p}\in B,\tilde{\eta}_n=y)=\int_B P_f^{\vartheta}(\eta_n=y)\mu(d\vartheta)$. Moreover we have $\tau_n(h,B)=W_f(\tilde{p}\in B,\tilde{\eta}_n=y)/\overline{P}_f(\eta_n=y)$. Now we get

$$\begin{aligned}\prod_o^n\overline{p}_{\nu f}(y_{\nu+1})&=\overline{P}_f(\eta_n=y)\overline{p}_{nf}(y,s)=\overline{P}_f(\eta_n=y)\int\vartheta(s_n,f_n(y),s)\tau_n(h,d\vartheta)=\\&=\int\vartheta(s_n,f_n(y),s)P_f^{\vartheta}(\eta_n=y)\mu(d\vartheta)\\&=\int P_f^{\vartheta}(\eta_{n+1}=(y,s))\mu(d\vartheta)=\overline{P}_f(\eta_{n+1}=(y,s)).\end{aligned}$$

+) We write here $\vartheta(i,a,j)$ instead of the map $\vartheta\to\vartheta(i,a,j)$.

Hence (10.11) is true for n+1, and the proof is complete. ┘

We shall denote by $\overline{DM}$ the DM $(S,A,D,(\overline{p}_n),r_n)$, to which the original DMU has been reduced, by theorem 10.3.

Since we have developped the theory of general DM's in previous sections, theorem 10.3 represents a powerful tool for dealing with our BDM. Many theorems for which complicated and sometimes unsatisfactory proofs are given in the literature, may now be easily deduced from theorems in previous sections. A sample of results obtainable in this way is given below in theorems 10.6. through 10.9.

Let ρ be either a probability measure on $\mathfrak{T}$ or the identically vanishing measure. To any $s\in S$, $a\in D(s)$, $j\in S$ we associate the measure $\Psi_{s,a,j}(\rho)$ on $\mathfrak{T}$, which assumes at the set B the value $\int_B \vartheta(s,a,j)\rho(d\vartheta)/\int\vartheta(s,a,j)\rho(d\vartheta)$.

Lemma 10.4. The a posteriori distributions of $\tilde{p}$ satisfy the recurrence relation

$$\tau_{n+1}(h,a,j,\cdot) = \Psi_{s_n,a,j}(\tau_n(h,\cdot)),\ n\in\mathbb{N},(h,a,j)\in H_{n+1}. \tag{10.12}$$

Proof. Let f be a plan in $\Delta_{n+1}(h,a,j)$, hence $f_n(y)=a$ and $\tau_{n+1}(h,a,j,B)=W_f(\tilde{p}\in B|\tilde{\eta}_{n+1}=(y,j))$. At first we assume $W_f(\tilde{\eta}_{n+1}=(y,j))=W_f(\tilde{\eta}_n=y)\cdot\overline{p}_{nf}(y,j)>0$. Then we get from (10.10)

$$\begin{aligned}\tau_{n+1}(h,a,j,B)&=\int_B\vartheta(s_n,a,j)\frac{P_f(\eta_n=y)\mu(d\vartheta)}{W_f(\eta_n=y)}\cdot\frac{1}{\int\vartheta(s_n,a,j)\tau_n(h,d\vartheta)} =\\ &=\int_B\vartheta(s_n,a,j)\tau_n(h,d\vartheta)/\int\vartheta(s_n,a,j)\tau_n(h,d\vartheta) =\\ &= \Psi_{s_n,a,j}(\tau_n(h,B)).\end{aligned}$$

Now let us assume $W_f(\tilde{\eta}_{n+1}=(y,j))=0$, hence $\tau_{n+1}(h,a,j,\cdot)\equiv 0$. If $W_f(\tilde{\eta}_n=y)=0$, then $\tau_n(h,\cdot)\equiv 0$ and therefore $\Psi_{s_n,a,j}(\tau_n(h,\cdot))\equiv 0$. If $\overline{p}_{nf}(y,j)=0$, then $\int\vartheta(s_n,a,j)\tau_n(h,d\vartheta)=\overline{p}_{nf}(y,j)=0$, hence also $\Psi_{s_n,a,j}(\tau_n(h,\cdot))\equiv 0$. ┘

Now we get

Theorem 10.5. The sequence (t_n), defined by

$$t_n(h):=(s_n,\tau_n(h,\cdot)),\ n\in\mathbb{N},h\in H_n, \tag{10.13}$$

is a sufficient statistic for the model $\overline{DM}$.

Proof. We have to verify the conditions (α) and (β) of section 6, i.e. we have to show: If $n\in\mathbb{N},h\in H_n,h'\in H_n,a\in D_n(h),s\in S$, and $t_n(h)=t_n(h')$, i.e. $s_n=s_n'$ and $\tau_n(h,\cdot)=\tau_n(h',\cdot)$, then

$$(\alpha) \qquad \begin{cases} D_n(h) = D_n(h'), \\ \overline{p}_n(h,a,s) = \overline{p}_n(h',a,s), \\ r_n(h,a) = r_n(h',a), \end{cases}$$

$$(\beta) \qquad t_{n+1}(h,a,s) = t_{n+1}(h',a,s).$$

By the assumption for the model DMU we get

$$D_n(h)=D(s_n)=D(s_n')=D_n(h'),$$

$$\overline{p}_n(h,a,s)=\int\vartheta(s_n,a,s)\tau_n(h,d\vartheta)=\int\vartheta(s_n',a,s)\tau_n(h',d\vartheta)$$

$$= \overline{p}_n(h',a,s), \text{ if } \tau_n(h,\cdot)\not\equiv 0, \text{ and}$$

$$\overline{p}_n(h,a,s)=\vartheta(s_n,a,s)=\vartheta(s_n',a,s)=\overline{p}_n(h',a,s), \text{ if } \tau_n(h,\cdot)\equiv 0,$$

$$r_n(h,a) = r_n(s_n,a) = r_n(s_n',a) = r_n(h',a).$$

From Lemma 10.4 follows

$$t_{n+1}(h,a,s) = (s_n,\Psi_{s_n,a,s}(\tau_n(h,\cdot)))=(s_n',\Psi_{s_n',a,s}(\tau_n(h',\cdot)))=$$

$$= t_{n+1}(h',a,s). \quad \rfloor$$

Let F_n be the set $\{(s_n,\tau_n(h,\cdot)):h\in H_n\}$. It is now obvious that the function $T_n:F_n\to F_{n+1}$ that relates t_{n+1} to t_n (cf.section 6) by means of $t_{n+1}(h,a,j)=T_n(t_n(h),a,j)$ is given by

$$(10.14) \qquad T_n((s,\rho),a,j):=(j,\Psi_{s,a,j}(\rho)).$$

The following theorem is a consequence of theorem 6.0.

<u>Theorem 10.6</u>. (cp. Martin (67),p.37). Assume case (EP).

(i) The maximal expected reward $\overline{G}_n$ of the model $\overline{DM}$ depends on h only by means of $(s_n,\tau_n(h,\cdot))$, i.e.

$$\overline{G}_n(h) = I_n(s_n,\tau_n(h,\cdot)), n\in\mathbb{N}, h\in H_n,$$

for some uniquely defined map $I_n:F_n\to\overline{\mathbb{R}}$.

(ii) (I_n) satisfies the OE

$$(10.15) \qquad I_n(s,\rho) =$$

$$= \sup_{a\in D(s)} \left[r_n(s,a)+\sum_j\left(\int\vartheta(s,a,j)\rho(d\vartheta)\right)\cdot I_{n+1}(j,\Psi_{s,a,j}(\rho))\right],$$

$$n\in\mathbb{N}, \ (s,\rho)\in F_n.$$

(iii) The sequence (I_n) is the smallest of those solutions (v_n) of the OE that satisfy $v_n\geq-\sum_n^\infty\|r_\nu^-\|, n\in\mathbb{N}$. In case (C) the sequence (I_n) is the unique solution (v_n) that satisfies $\|v_n\|\leq\sum_n^\infty\|r_\nu\|, n\in\mathbb{N}$.

Put $F:=\cup F_n$ and let N be the set of functions from F to $\overline{\mathbb{R}}$ that are bounded from below. We define an operator $X_n:N \to N$ by means of

$$(10.16) \qquad (X_n v)(s,\rho) := \sup_{a \in D(s)} [r_n(s,a) + \sum_j (\int \vartheta(s,a,j)\rho(d\vartheta)) v(j, \Psi_{s,a,j}(\rho))],$$
$$n \in \mathbb{N},\ (s,\rho) \in F.$$

Then the OE (10.15) reads

$$I_n(s,\rho) = (X_n I_{n+1})(s,\rho), n \in \mathbb{N}, (s,\rho) \in F_n.$$

The following theorems constitute generalizations of results of Martin (67).

<u>Theorem 10.7.</u> (cf.theorem 4.1). Define the double sequence $(I_{nk}, n \in \mathbb{N}, k \in \mathbb{N})$ of functions $I_{nk}:F \to \overline{\mathbb{R}}$ by the recursion

$$I_{no} :\equiv 0,\ n \in \mathbb{N}, \text{ and}$$

$$(10.17) \qquad I_{nk} := X_n I_{n+1,k-1},\ n \in \mathbb{N}, k \in \mathbb{N}.$$

Then $I_n(s,\rho) = \lim_{k \to \infty} I_{nk}(s,\rho), n \in \mathbb{N}$, $(s,\rho) \in F_n$, if case (EP) holds.

<u>Theorem 10.8.</u> (cf.corollary 5.4). If each of the sets $D(s), s \in S$, is finite, then there exists in case (EN) a $\overline{p}$-optimal plan for the BDM.

<u>Remark.</u> (cf. remark 2 in section 6.) It can be shown that there exists a $\overline{p}$-optimal plan f depending only on τ_n in the following sense:

$$(10.18) \qquad \tau_n(h_{nf}(y),\cdot) = \tau_n(h_{nf}(y'),\cdot) \text{ and}$$
$$s_n = s'_n \text{ imply } f_n(y) = f_n(y').$$

By going back to the definition of τ_n, we may express property (10.18) also by the statement

$$W_f(\tilde{p} \in B \mid \tilde{\eta}_n = y) = W_f(\tilde{p} \in B \mid \tilde{\eta}_n = y')\ \forall B \in \mathfrak{F} \text{ and } s_n = s'_n$$

imply $f_n(y) = f_n(y')$.

Now we specialize to the case of stationary BDM, i.e. to a BDM where we have $r_n = \beta^{n-1} r$ for some function r and some $\beta \in (0,1\rangle$.

<u>Theorem 10.9.</u> (cf.theorem 6.7.) Let the BDM be stationary and assume that either $r \geq 0$ or r bounded from below and $\beta < 1$.

Define the sequence (J_k) of functions $J_k: F \to \overline{\mathbb{R}}$ by

$$J_o := 0$$

(10.19) $$J_k(s,\rho) := $$

$$= \sup_{a \in D(s)} \left[r(s,a) + \beta \sum_j \left(\int \vartheta(s,a,j)\rho(d\vartheta) \right) J_{k-1}(j, \Psi_{s,a,j}(\rho)) \right],$$

$$= X_1(\beta J_{k-1}).$$

Then we have:

(i) $I := \lim_k J_k$ exists on F and $I_n = \beta^{n-1} I$ on F_n, $n \in \mathbb{N}$.

(ii) I is a solution of the OE

(10.20) $$I = X_1(\beta I).$$

If $r \geq 0$ then I is the smallest positive solution of (10.20). If r is bounded from below and $\beta < 1$, then I is the smallest of those solutions v of (10.20) for which $v > -\|r^-\|(1-\beta)^{-1}$. If r is bounded and $\beta < 1$ then I is the unique bounded solution of (10.20).

Proof. Since $r_n = \beta^{n-1} r$, we realize that the operator X_n, defined by (10.16), satisfies $X_n v = \beta^k X_{n-k}(\beta^{-k} v)$, $1 \leq k < n < \infty$, $v \in N$. From (10.17) we conclude by induction on k that $I_{nk} = \beta^{n-1} I_{1k} = \beta^{n-1} J_k$ holds on F. Theorem 10.7 implies that I exists on F and that $I_n = \beta^{n-1} I$ on F_n, whereas theorem 10.6(ii) implies that I satisfies the OE (10.20). The stated characterization of I among the solutions of the OE is shown as part (iii) of theorem 6.7. ⌋

Chapter II. General state space

11. Decision models.

If the state space S of a decision problem is not countable we have to take into account that in general we cannot take the system of all subsets of S as the domain of the probability measures we are interested in and that consequently the measurability of functions defined on S is no longer a triviality.

This leads to the following

Definition. A (general) *decision model* (DM) is a tupel $((S,\mathfrak{S}),(A,\mathfrak{A}),D,(q_n),(r_n))$ of the following meaning.

(i) S is a non-empty set, the so-called *state space*, endowed with a σ-algebra $\mathfrak{S}$.

(ii) A is a non-empty set, the so-called *space of actions*, endowed with a σ-algebra $\mathfrak{A}$. As earlier, $\overline{H}_n$ will denote the set $S\times A\times S\times\ldots\times S$ (2n-1 factors) of *histories* $h_n=(s_1,a_1,\ldots,s_n)$. Moreover, $\overline{\mathfrak{H}}_n$ will denote the product-σ-algebra $\mathfrak{S}\otimes\mathfrak{A}\otimes\ldots\otimes\mathfrak{S}$ in $\overline{H}_n$.

(iii) D is a sequence of maps D_n from sets $H_n\subset\overline{H}_n$ to the set of all non-empty subsets of A with the property that

$$H_1 := S \quad \text{and that}$$

$H_{n+1}:=\{(h,a,s)\in\overline{H}_{n+1}:\ h\in H_n, a\in D_n(h), s\in S\}$ belongs to $\overline{\mathfrak{H}}_{n+1}$ for all $n\in\mathbb{N}$. $D_n(h)$ is called the *set of admissible actions* at time n under history h. The trace-σ-algebra $H_n\cap\overline{\mathfrak{H}}_n$ will be denoted by $\mathfrak{H}_n$ and the set $\{(h,a):h\in H_n, a\in D_n(h)\}$ will be denoted by K_n. We assume that K_n contains the graph of a measurable map.

(iv) q_o is a probability [+)] on $\mathfrak{S}$, the so-called *initial distribution*. We shall often write p instead of q_o. $q_n(h,a,\cdot)$ is a transition probability from $(K_n,K_n\cap(\mathfrak{H}_n\otimes\mathfrak{A}_n))$ to $(S,\mathfrak{S})$, the so-called *transition law* between time n and n+1.

(v) r_n is an extended real valued measurable [++)] function on K_n, the so-called *reward* during the time interval $(n,n+1>$.

+) Instead of a 'probability measure' we shall speak only of a 'probability'.

++) We shall call a set or a map measurable without mentioning the σ-algebras involved whenever it is clear from the context which σ-algebras are meant. In particular, if B is a non-empty subset in a measurable space $(F,\mathfrak{F})$, then B will be endowed with the trace-σ-algebra $B\cap\mathfrak{F}$; and if $(F_i,\mathfrak{F}_i)$, $i\in I$, is a family of measurable spaces, then $\times_{i\in I}F_i$ will be endowed with the product-σ-algebra.

Remark. K_n is the s-section of the set $H_{n+1} \in \overline{\mathfrak{H}}_{n+1}$ for any s, hence K_n belongs to $\overline{\mathfrak{H}}_n \otimes \mathfrak{A}$. Furthermore we have $D_n(h)=(K_n)_h$ for any $h \in H_n$, hence $D_n(h)$ belongs to $\mathfrak{A}$.

Definition. a) A (deterministic admissible) *plan* is a sequence $f=(f_n)$ of measurable maps $f_n : S^n \to A$ with the property

$$(11.1) \qquad f_n(y) \in D_n(h_{nf}(y)), \quad n \in \mathbb{N}, \quad y \in S^n,$$

where

$$(11.2) \qquad h_{nf}(y) := (s_1, f_1(s_1), s_2, f_2(s_1, s_2), \ldots \ldots, f_{n-1}(s_1, \ldots, s_{n-1}), s_n)$$

denotes the history at time n obtained by the use of f when the sequence $y := (s_1, s_2, \ldots, s_n)$ of states occurred.

b) A *randomized* (admissible) *plan* is a sequence $\pi=(\pi_n)$ of transition probabilities π_n from $(H_n, \mathfrak{H}_n)$ to $(A, \mathfrak{A})$ such that

$$(11.3) \qquad \pi_n(h, D_n(h)) = 1, \quad n \in \mathbb{N}, \quad h \in H_n.$$

Let Δ and Δ^r be the (non-empty)+) sets of plans and randomized plans, respectively. Whenever we are dealing with (deterministic) plans we shall assume that the σ-algebra contains all singletons (sets consisting of one point). This assumption is fulfilled e.g., if $\mathfrak{A}$ is generated by a Hausdorff-topology, in particular if $(A, \mathfrak{A})$ is a standard Borel space (cf. section 12).

Let π be a randomized plan, such that each of the probabilities $\pi_n(h, \cdot)$ is concentrated at some point $g_n(h) \in A$, i.e. such that $\pi_n(h,B)=1_B(g_n(h))$, $n \in \mathbb{N}, h \in H_n, B \in \mathfrak{A}$, for some map $g_n : H_n \to A$. In that case one will interpret π as a deterministic plan g for which g_n does not only depend on the state history but on the whole history at time n. Let us call g an *H-plan* and let Δ' be the set of all H-plans.

Obviously Δ' is the set of all sequences (g_n) of maps $g_n : H_n \to A$, such that g_n is measurable and such that

$$(11.4) \qquad g_n(h) \in D_n(h), \quad n \in \mathbb{N}, h \in H_n.$$

An H-plan g and the corresponding plan $\pi \in \Delta^r$ determine each other uniquely since $\mathfrak{A}$ contains all singletons. Hence we shall regard Δ' as a subset of Δ^r. We shall show in lemma

+) Since K_n contains the graph of a measurable map $g_n : H_n \to A$, the sequence (f_n), determined by (11.5), belongs to Δ.

15.3. that the map $g \to f$, defined by

(11.5) $$f_n(y) := g_n(h_{nf}(y)), \; n \in \mathbb{N}, \; y \in S^n,$$

is a map from Δ' *onto* Δ. It should also be noted that for any $f \in \Delta$ the map $h_{nf}: S^n \to H_n$ is measurable.

Now we have to construct *probability spaces* $(\Omega, \mathfrak{F}, P_f)$ and $(\overline{H}, \mathfrak{H}, P_\pi)$, determined by an arbitrary plan $f \in \Delta$ and an arbitrary plan $\pi \in \Delta^r$, respectively. As in sections 2 and 9 we shall make the following choice for this construction:

a) $\Omega := S^{\mathbb{N}}$; $\mathfrak{F} := \bigotimes_1^\infty \mathcal{S}$. Let q_{nf} be the transition probability from S^n to S, defined by

(11.6) $$q_{nf}(y, \cdot) := q_n(h_{nf}(y), f_n(y), \cdot), \; n \in \mathbb{N}, y \in S^n.$$

Let ζ_n denote the n-th coordinate variable on $(\Omega, \mathfrak{F})$, and put $\eta_n := (\zeta_1, \zeta_2, \ldots, \zeta_n)$. Then, according to a theorem of C. Ionescu Tulcea (cf. theorem A6), there exists a unique probability measure P_f on $\mathfrak{F}$ such that

(11.7) $$P_f(\eta_n \in \times_1^n B_\nu) = \int_{B_1} p(ds_1) \int_{B_2} q_{1f}(s_1, ds_2) \ldots$$

$$\ldots \int_{B_n} q_{n-1,f}(y_{n-1}, ds_n), \; n \in \mathbb{N}, B_\nu \in \mathcal{S}.$$

P_f has the property that

$$E_f \varphi \circ \eta_n = \int p(ds_1) \int q_{1f}(s_1, ds_2) \ldots \int q_{n-1,f}(y_{n-1}, ds_n) \varphi(y_n)$$

$$=: p q_{1f} \cdots q_{n-1,f} \varphi$$

holds whenever $\varphi: S^n \to \overline{\mathbb{R}}$ is measurable and $\varphi \circ \eta_n$ is quasi-integrable with respect to P_f.

b) $\overline{H} := S \times A \times S \ldots$, $\mathfrak{H} := \mathcal{S} \otimes \mathfrak{A} \otimes \mathcal{S} \otimes \ldots$. Let ζ_n +) and α_n be the projection from $\overline{H}$ into the n-th state space and n-th action space, respectively. Then $\chi_n := (\zeta_1, \alpha_1, \zeta_2, \ldots, \zeta_n)$ describes the history at time n. In order to be able to define a probability measure on $\mathfrak{H}$ by means of (q_n) and π, we have to extend the definition of q_n and π_n from $K_n \times \mathcal{S}$ and $H_n \times \mathfrak{A}$ to $(\overline{H}_n \times A) \times \mathcal{S}$ and $\overline{H}_n \times \mathfrak{A}$, respectively. This can be done in many different ways. Condition (11.3) implies that for any extension there exists a unique probability measure P_π (independent of the extension) on $\mathfrak{H}$ such that

---

+) From the context it will always be clear whether ζ_n is defined on Ω or on $\overline{H}$.

(11.8) $$P_\pi(\chi_n \in B_1\times C_1\times\ldots\times B_n) = \int_{B_1} p(ds_1)\int_{C_1} \pi_1(s_1,da_1)\int_{B_2} q_1(s_1,a_1,ds_2)\ldots \int_{B_n} q_{n-1}(h_{n-1},a_{n-1},ds_n),\ n\in\mathbb{N}, B_\nu\in\mathfrak{S}, C_\nu\in\mathfrak{A}.$$

It follows that

$$E_\pi\Psi\circ\chi_n = \int p(ds_1)\int\pi_1(s_1,da_1)\ldots\int q_{n-1}(h_{n-1},a_{n-1},ds_n)\Psi(h_n) =: p\pi_1\ldots q_{n-1}\Psi$$

holds whenever $\Psi:\overline{H}_n\to\overline{\mathbb{R}}$ is measurable and whenever $\Psi\circ\chi_n$ is quasi-integrable with respect to P_π. One easily proves that $P_\pi(\chi_n\in H_n)=1$, that $H:=\{(s_1,a_1,\ldots)\in\overline{H}: a_\nu\in D_\nu(h_\nu),\ \nu\in\mathbb{N}\} = \bigcap_n(H_n\times A\times S\times\ldots)$ belongs to $\overline{\mathfrak{H}}$, and that $P_\pi(H)=1$.

If we use plan $f\in\Delta$ and if the sequence $y\in S^n$ of states has occurred then we receive during time period $(n,n+1\rangle$ the reward

$$r_{nf}(y) := r_n(h_{nf}(y),f_n(y)).$$

Under either of the assumptions (EN) and (EP) of section 2 there is defined the total reward

$$R_f := \sum r_{nf}\circ\eta_n$$

as an extended real valued and quasi-integrable random variable on $(\Omega,\mathfrak{F},P_f)$. Hence there exists the *expected total reward*

$$G_f := E_fR_f = \int R_f dP_f.$$

As in section 2 we get

(11.9) $$-\infty\leq G_f\leq\sum\|r_n^+\|<\infty \quad \text{in case (EN)},$$

(11.10) $$-\infty<-\sum\|r_n^-\|\leq G_f\leq\infty \quad \text{in case (EP)},$$

(11.11) $$|G_f|\leq\sum\|r_n\|<\infty \quad \text{in case (C)}.$$

Definition. The plan $f^*\in\Delta$ is called *$\overline{p}$-optimal* if

$$G_{f^*} = \sup_{f\in\Delta} G_f =: G.$$

Now let us turn to the case of a randomized plan $\pi\in\Delta^r$. In order to be able to define $r_n\circ(\chi_n,\alpha_n)$ everywhere on $\overline{H}$, we have to extend the definition of r_n from K_n to $\overline{H}_n\times A$. For any measurable extension there exists under either of the assumptions (EN) and (EP) P_π-a.s. the *total reward*

$$R := \sum r_n \circ (\chi_n, \alpha_n)$$

as an extended real valued and quasi-integrable random variable on $(\overline{H}, \overline{\mathfrak{H}}, P_\pi)$. In order to have R defined everywhere on $\overline{H}$ we define $r_n(h,a) := 0$ for $(h,a) \in \overline{H}_n \times A - K_n$. The *expected total reward* is given by

$$G_\pi := E_\pi R = \int R dP_\pi .$$

Of course, the inequalities (11.9)-(11.11) remain true if f is replaced by π.

As in chapter I we have to consider the *conditional expected reward* for the time period (n,∞) under the use of f:

$$(11.12) \qquad G_{nf}(y) := E_f\Big[\sum_{i=n}^{\infty} r_{if} \circ \eta_i \,|\, \eta_n = y\Big], \; n \in \mathbb{N}, y \in S^n, f \in \Delta,$$

and also the *conditional expected reward* under the use of π:

$$(11.13) \qquad V_{n\pi}(h) := E_\pi\Big[\sum_{i=n}^{\infty} r_i \circ (\chi_i, \alpha_i) \,|\, \chi_n = h\Big],$$

$$n \in \mathbb{N}, h \in H_n, \pi \in \Delta^r .$$

According to our definition of conditional expectations by means of the given conditional distributions (cf. appendix 3) G_{nf} and $V_{n\pi}$ are uniquely defined, and moreover

$$(11.14) \qquad -\infty \le G_{nf} \le \sum_{i=n}^{\infty} \|r_i^+\| < \infty \qquad \text{in case (EN)},$$

$$(11.15) \qquad -\infty < -\sum_{i=n}^{\infty} \|r_i^-\| \le G_{nf} \le \infty \qquad \text{in case (EP)},$$

$$(11.16) \qquad |G_{nf}| \le \sum_{i=n}^{\infty} \|r_i\| < \infty \qquad \text{in case (C)}.$$

The equations (11.14)-(11.16) remain valid when G_{nf} is replaced by $V_{n\pi}$.

As in section 3 we shall regard the set

$$\Delta_n(h) := \{f \in \Delta:\ f_\nu(y_\nu) = a_\nu, 1 \le \nu < n\}$$

of plans that are available at time n, when history h has occurred. Now we can define the *maximal conditional expected reward*

$$G_n(h) := \sup_{f \in \Delta_n(h)} G_{nf}(y), \; n \in \mathbb{N}, \; h \in H_n,$$

and $\qquad V_n(h) := \sup_{\pi \in \Delta^r} V_{n\pi}(h), \; n \in \mathbb{N}, h \in H_n.$

According to our definition of $r_n(h,a):=0$ for $(h,a)\notin K_n$, $V_{n\pi}(h)$ and $V_n(h)$ are also defined for $h\notin K_n$ and equal zero. This fact will only play a minor role in some proofs.

Lemma 3.5 and lemma 3.6 carry over immediately to

Lemma 11.1. For any $n\in\mathbb{N}$ and $f\in\Delta$ we have

(11.17) $$G_f = \int G_{1f}dp,$$

(11.18) $$G_{nf} = r_{nf} + \int G_{n+1,f}(\cdot,s)q_{nf}(\cdot,ds) =: \Lambda_{nf}G_{n+1,f},$$

(11.19) $$G_{nf} = \lim_k \Lambda_{nf}\Lambda_{n+1,f}\cdots\Lambda_{n+k,f}0.$$

When we are dealing with randomized plans, the role of the operator Λ_{nf} is taken over by the operator $\Lambda_{n\pi}$, defined by

(11.20) $$(\Lambda_{n\pi}u)(h):=\int\pi_n(h,da)\left[r_n(h,a)+\int q_n(h,a,ds)u(h,a,s)\right], \quad n\in\mathbb{N}, h\in H_n.$$

Let M_n^+ and M_n^- be the set of measurable functions $v:H_n\to\overline{\mathbb{R}}$ that are bounded from below in case (EP) and bounded from above in case (EN), respectively. Then $\Lambda_{n\pi}$ is a map from M_{n+1}^+ into M_n^+ in case (EP) and from M_{n+1}^- into M_n in case (EN).

Lemma 11.1 carries over to

Lemma 11.2. For any $n\in\mathbb{N}$ and $\pi\in\Delta^r$ we have

(11.21) $$G_\pi = \int V_{1\pi}dp,$$

(11.22) $$V_{n\pi} = \Lambda_{n\pi}V_{n+1,\pi},$$

(11.23) $$V_{n\pi} = \lim_k \Lambda_{n\pi}\Lambda_{n+1,\pi}\cdots\Lambda_{n+k,\pi}0.$$

12. Measure-theoretic and topological preparations.

In this section we collect some prerequisites which are necessary for the development of the theory in the subsequent sections. One of our goals will be to show that under certain assumptions the sequence (V_n) of maximal conditional expected rewards is a solution (w_n) of the *optimality equation* (OE)

$$(12.1) \qquad w_n(h) := \sup_{a\in D_n(h)} \left[r_n(h,a)+\int q_n(h,a,ds)w_{n+1}(h,a,s)\right],$$

$$=: U_n w_{n+1}(h),\ n\in \mathbb{N},\ h\in H_n.$$

There have to be made some remarks concerning the statement of the OE. The right hand side of (12.1) is defined whenever w_{n+1} belongs to M^-_{n+1} or M^+_{n+1} in case (EN) and case (EP), respectively. However, since V_n is the supremum of the measurable functions $G_{n\pi}$, taken over the (in general) not countable set Δ^r, we do not know in advance that V_n is $\mathfrak{H}_n$-measurable. In fact, Strauch (66) has given an example in which V_n is *not* $\mathfrak{H}_n$-measurable. We shall reproduce his example after the proof of theorem 14.5. On the other hand, we shall show in section 13, generalizing a result of Strauch (66), that V_n is *universally* measurable in the last coordinate which is sufficient for the existence of the right hand side of (12.1).

<u>Definition.</u> Let $(X,\mathfrak{L})$ be a measurable space and W the set of probability measures on $\mathfrak{L}$. Denote for any $w\in W$ the completion of $(X,\mathfrak{L},w)$ by $(X,\mathfrak{L}_w,\overline{w})$. Then $\hat{\mathfrak{L}}:=\bigcap_{w\in W}\mathfrak{L}_w$ is called the σ-algebra of $\mathfrak{L}$-*universally measurable* (abbreviated by u-measurable) sets.

Universal measurability of sets and functions will mean measurability with respect to $\hat{\mathfrak{L}}$. Measurability implies u-measurability since $\mathfrak{L}\subset\hat{\mathfrak{L}}$. By the definition of $\hat{\mathfrak{L}}$ any probability measure w on $\mathfrak{L}$ has a unique extension w' to $\hat{\mathfrak{L}}$. In the sequel we shall simply write w instead of w'.

<u>Lemma 12.1.</u> Let $(X,\mathfrak{L},\mu)$ be a probability space. Let $u:X\to\overline{\mathbb{R}}$ be $\mathfrak{L}$-universally measurable. Then there exists a set $N\in\mathfrak{L}$ with $\mu(N)=0$ and a $\mathfrak{L}$-measurable map $v:X\to\overline{\mathbb{R}}$ such that $u=v$ on N^c.

Proof. At first we assume that u is the indicator of some set $F \in \hat{\mathcal{L}}$, $u := 1_F$. $\hat{\mathcal{L}}$ is a subsystem of the completion $\mathcal{L}_\mu$, hence $F \in \mathcal{L}_\mu$. Then there exist a partition $F = F' + N'$, where F' belongs to $\mathcal{L}$ and where N' is subset of a set $N \in \mathcal{L}$ of μ-measure zero. Now $v := 1_{F'}$ has the desired property. The proof for arbitrary $\hat{\mathcal{L}}$-measurable u follows by a standard argument. $\rfloor$

<u>Definition</u>. Let $(X, \mathcal{L})$ be a measurable space and M an arbitrary non-empty subset of the set of finite measures on $\mathcal{L}$. Then the smallest of those σ-algebras $\mathcal{G}$ in M for which any of the maps $L_B : M \to \mathbb{R}, B \in \mathcal{L}$, defined by $L_B(\mu) := \mu(B), \mu \in M$, is $\mathcal{G}$-$\mathcal{L}_1$+) -measurable, will be called the **-σ-algebra* $\mathcal{M}$ in M.

<u>Remarks</u>. (i) By definition, $\mathcal{M}$ is generated by the system of sets $(\{\mu \in M : \mu(B) \in F\}, B \in \mathcal{L}, F \in \mathcal{L}_1)$.

(ii) If M' is a non-empty subset of M, then $M' \cap \mathcal{M}$ is the *-σ-algebra in M'. This fact follows from (i) by means of the following well-known result (cf.Neveu (65),p.19):
Let $(X, \mathcal{F})$ be a measurable space and Y a non-empty subset of X. If $\mathcal{F}$ is generated by the system $\mathcal{E}$, then $Y \cap \mathcal{F}$ is generated by $Y \cap \mathcal{E}$.

(iii) The usefulness of the concept of a *-σ-algebra becomes already apparent in the following fact: Let M be a set of probability measures on $\mathcal{L}$ with *-σ-algebra $\mathcal{M}$. Let $(Z, \mathfrak{z})$ be a measurable space and φ a map from Z into M. Then φ is $\mathfrak{z}$-$\mathcal{M}$-measurable iff the map $(z,B) \to \varphi(z)(B)$, $(z,b) \in Z \times \mathcal{L}$, is a transition probability from $(Z, \mathfrak{z})$ to $(X, \mathcal{L})$.

<u>Lemma 12.2</u>.(cf.Strauch (66),lemma 7.1).
Let $(Z, \mathfrak{z})$ and $(X, \mathcal{L})$ be measurable spaces, let $\mathcal{W}$ be the *-σ-algebra in the set W of probabilities on $\mathcal{L}$. Let $u : Z \times X \to \overline{\mathbb{R}}$ be $\mathfrak{z} \otimes \mathcal{L}$-measurable and either bounded from above or from below. Then the map

$$(12.2) \qquad (z, \mu) \to \int u(z, \cdot) d\mu$$

is $\mathfrak{z} \otimes \mathcal{W}$-measurable.

The proof given in Strauch (66) refers to a paper of Dubins/Freedmann (64). This reference seems to be not quite clear, therefore we give here a complete proof. As in lemma

---

+) $\mathcal{L}_1$ is the σ-algebra of Borel sets in $\mathbb{R}$.

12.1, we need only consider the case where u is the indicator 1_F of a set $F \in \mathcal{Z}\otimes\mathcal{B}$. In that case $\int u(z,\cdot)d\mu$ reduces to $\mu(F_z)$, where F_z is the z-section of F. We define $\mathcal{D}:=\{C \in \mathcal{Z}\otimes\mathcal{B}: (z,\mu)\to\mu(C_z)$ is $\mathcal{Z}\otimes\mathcal{M}$-measurable$\}$. One easily checks that $\mathcal{D}$ is a Dynkin-system (cf.Bauer (68)), which contains the system $\mathcal{Z}\times\mathcal{B}$ of rectangles with sides in $\mathcal{Z}$ and $\mathcal{B}$. Since $\mathcal{Z}\times\mathcal{B}$ is closed under the intersection of two sets, we conclude (cf.Bauer (68),p.18) that $\mathcal{D}$ equals $\mathcal{Z}\otimes\mathcal{B}$, which is the desired result. ⌋

One easily deduces from lemma 12.2. the

Corollary 12.3. Let $(Z,\mathcal{Z})$, $(X,\mathcal{B})$ and $\mathcal{M}$ be as in lemma 12.2. Let Y be a measurable map from X into a measurable space $(\Gamma,\mathcal{G})$, let $v:Z\times\Gamma\to\overline{\mathbb{R}}$ be measurable and either bounded from below or from above. Let μ_Y be the image of the probability μ on $\mathcal{B}$ under Y. Then the map

$$(z,\mu) \to \int v(z,\cdot)d\mu_Y$$

is $\mathcal{Z}\otimes\mathcal{M}$-measurable.

Definition. A complete separable metric space is called a *Polish space*. We shall call the topology induced by a metric in a Polish space a *Polish topology*.

Remarks. We shall state some properties of Polish spaces the proofs of which may be found e.g. in Bourbaki (58).

(i) Any compact metric space is a Polish space. Any locally compact topological space with countable base can be endowed with a metric such that it becomes a Polish space.

(ii) Any countable set is a Polish space under the metric $d(x,y):=\delta_{xy}$.

(iii) A countable (topological) product of Polish spaces is a Polish space.

(iv) A subspace F of a Polish space E is a Polish space iff F is the intersection of a countable family of open subsets of E. In particular, any open and any closed subset of a Polish space is a Polish space.

(v) $\mathbb{R}^d$ is obviously a Polish space under the usual topology. It follows from (iv) that any closed and any open subset of $\mathbb{R}^n$ is a Polish space.

(vi) A Banach space is not necessarily a Polish space. E.g. the Banach space of bounded sequences of real numbers (which has no countable base) is not a Polish space.

Definition. Let $(E,\mathcal{T})$ be a topological space. Then the σ-algebra $\sigma(\mathcal{T})$ (the σ-algebra generated by the topology $\mathcal{T}$) is called the system of *Borel sets* in $(E,\mathcal{T})$.

In the theory of probability it has turned out that general measurable spaces $(E,\mathcal{L})$ fail to have some of the properties (e.g. existence of conditional distributions) one would expect from any reasonable model for reality. But fortunately it also turned out that a measurable space $(E,\mathcal{L})$ has nice properties if $\mathcal{L}$ is generated by a Polish topology on E. This condition may often be weakened to the condition that $(E,\mathcal{L})$ is a so-called standard-Borel space.

Definition. A measurable space $(E,\mathcal{L})$ is called a *standard Borel space* (SB-space) if E belongs to the system $\mathcal{G}$ of Borel sets in a Polish space and if $\mathcal{L}$ is the trace of $\mathcal{G}$ on E.

Remarks.(i) It is easy to show that if F is a non-empty measurable set in an SB-space $(E,\mathcal{L})$ then $(F,F\cap\mathcal{L})$ is an SB-space.
(ii) In an SB-space all singletons are measurable.

Theorem 12.4. Let $(M,\mathcal{M},P)$ be a probability space, $(\Omega_1,\mathcal{A}_1)$ a measurable space and $(\Omega_2,\mathcal{A}_2)$ a standard Borel space. Let $X_i:M\to\Omega_i$ be measurable maps, i=1,2. Then there exists a conditional distribution of X_2 under the condition X_1.

Theorem 12.4 is well-known, but usually its proof is given only for the case where $\mathcal{A}_2$ is generated by a Polish topology. We shall sketch how the general case may be derived from a theorem that is more easily found in the literature.
α) Let $(\Omega,\mathcal{A},\mu)$ be a probability space, $\mathcal{G}$ a sub-σ- algebra of $\mathcal{A}$, E a Polish space with topology $\mathcal{T}$, $X:\Omega\to E$ $\mathcal{A}$-$\sigma(\mathcal{T})$-measurable Then there exists (cf.e.g. Bauer (68),p.258) a transition probability q from $(\Omega,\mathcal{G})$ into $(E,\sigma(\mathcal{T}))$ such that

$$\mu(G\cap[X\in F])=\int_G \mu(d\omega)q(\omega,F),\quad G\in\mathcal{G}, F\in\sigma(\mathcal{T}). \tag{12.3}$$

(q is called a conditional distribution of X under the condition $\mathcal{G}$.)
β) We assert that one may replace in α) the measurable space $(E,\sigma(\mathcal{T}))$ by an SB-space $(\Phi,\mathcal{F})$. In fact, assume that $\mathcal{T}$ is a

Polish topology in E such that $\Phi\in\sigma(\mathcal{T})$ and $\mathcal{F}=\Phi\cap\sigma(\mathcal{T})$. Define $Y:\Omega\to E$ by $Y(\omega):=X(\omega),\omega\in\Omega$, hence Y is $\mathfrak{A}$-$\sigma(\mathcal{T})$-measurable. Let q be as in α) and put $Q(\omega,F):=q(\omega,F),F\in\mathcal{F}$. It follows easily that $Q(\cdot,F)$ is $\mathcal{G}$-measurable for any $F\in\mathcal{F}$ and that $Q(\omega,\cdot)$ is a measure on $\mathcal{F}$ for any $\omega\in\Omega$ such that $\int Q(\omega,\Phi)\mu(d\omega)=1$. Hence there exists a set $N\in\mathcal{G}$ of μ-measure zero such that $Q(\omega,\Phi)=1$ for $\omega\notin N$. If we replace $Q(\omega,\cdot)$ on N by an arbitrary probability measure on $\mathcal{F}$, then we get a transition probability with the desired properties.

γ) In order to prove theorem 12.4 we put $\Omega:=\Omega_1\times\Omega_2$, $\mathfrak{A}:=\mathfrak{A}_1\otimes\mathfrak{A}_2$, $\mu:=$ joint distribution of (X_1,X_2) under P, $\mathcal{G}:=\{B_1\times\Omega_2:B_1\in\mathfrak{A}_1\}$ = inverse image of $\mathfrak{A}_1$ under the projection $p_1:\Omega\to\Omega_1$, $X:=$projection $p_2:\Omega\to\Omega_2$. $(\Phi,\mathcal{F}):=(\Omega_2,\mathfrak{A}_2)$. Then there exists according to part β) a transition probability q from $(\Omega_1\times\Omega_2,\mathcal{G})$ into $(\Omega_2,\mathfrak{A}_2)$ such that, according to (12.3),

$$P(X_1\in B_1,X_2\in B_2)=\int_{B_1}P_{(X_1,X_2)}(d(x_1,x_2))q((x_1,x_2),B_2),\qquad B_i\in\mathfrak{A}_i .$$

$q(\cdot,B_2)$ does not depend on x_2, since it is $\mathcal{G}$-measurable. If we put $q'(x_1,\cdot):=q((x_1,x_2),\cdot)$ for arbitrary x_2, then q' is a conditional distribution of X_2 under X_1. ⌋

Corollary 12.5. Let $(\Omega_i,\mathcal{F}_i)$ be SB-spaces and let P be a probability measure on $\mathcal{F}:=\bigotimes_1^\infty\mathcal{F}_i$. Let P_o be the distribution of the projection from $\bigtimes_1^\infty\Omega_i$ into Ω_1. Then there exist transition probabilities P_o from $\bigtimes_1^i\Omega_i$ into Ω_{i+1} such that $P=\bigotimes_o^\infty P_i$.

Proof. Let X_i be the projection from $\bigtimes_1^\infty\Omega_i$ into Ω_i. According to theorem 12.4 there exists for any $i\in\mathbb{N}$ a conditional distribution of X_{i+1} under the condition $(X_1,X_2,\dots,X_i)$. In order to prove that $P=\bigotimes_o^\infty P_i$ it is sufficient to show that $\int g\circ(X_1,\dots,X_n)dP=\int P_o(dx_1)\dots\int P_{n-1}(x_1,\dots,x_{n-1},dx_n)\ (x_1,\dots,x_n)$ holds for any measurable and non-negative map $g:\bigtimes_1^n\Omega_i\to\overline{\mathbb{R}}$. This fact follows easily by induction on n. ⌋

Lemma 12.6 (Blackwell and Ryll-Nardzewski (63)). Let $(X,\mathcal{F})$ and $(Y,\mathcal{G})$ be SB-spaces, let B be a set in $\mathcal{F}\otimes\mathcal{G}$, let q be a transition probability from $(X,\mathcal{F})$ to $(Y,\mathcal{G})$. Then there exists a measurable map $f:X\to Y$ such that

(12.4) $\quad f(x)\in B_x$ whenever $q(x,B_x)>0$.

Remarks. (i) The condition $q(x,B_x)>0$ in (12.4) cannot be weakened to $B_x\neq\emptyset$. Literature on this problem can be found in Blackwell and Ryll-Nardzewski (63). See also lemma 12.12. (ii) Hinderer (67) has shown that one can choose f in such a way that in addition to (12.4) f(x) belongs to the support of the probability measure $q(x,\cdot)$ whenever $q(x,B_x)>0$. (iii) Lemma 12.6 has been proved by Blackwell and Ryll-Nardzewski for the special case that $q(x,B_x)>0$ for all $x\in\Phi$. Our formulation may be easily reduced to this special case. (iv) In the definition of our DM in section 11 we made the assumption that each of the sets K_n contains the graph of a measurable map. Obviously this assumption is equivalent to the assumption that the set Δ of plans is not empty. It follows from Lemma 12.6 that it is also equivalent to the assumption that the set Δ^r of randomized plans is not empty.

Corollary 12.7. Let $(X,\mathfrak{F})$, $(Y,\mathfrak{G})$, B and q be as in lemma 12.6. Let μ be a probability on $\mathfrak{F}\otimes\mathfrak{G}$ concentrated on B. If $q(x,B_x)>0$ for all $x\in pr_1(B)$, then there exists a conditional distribution $Q\in\mu_{pr_2|pr_1}$ such that $Q(x,B_x)=1$ for all $x\in pr_1(B)$.

Proof. According to theorem 12.4 there exists an element $q'\in\mu_{pr_2|pr_1}$. From $1=\mu(B)=\int\mu_{pr_1}(dx)q'(x,B_x)$ we get $q'(x,B_x)=1$ for all x in the complement N^c of a set $N\in\mathfrak{F}$ of μ_{pr_1}-measure zero. From lemma 12.6 we conclude the existence of a (degenerate) transition probability q" from X to Y such that $q''(x,B_x)=1, x\in pr_1(B)$. The transition probability $Q(x,\cdot):=q'(x,\cdot)1_{N^c}(x)+q''(x,\cdot)1_N(x), x\in X$, has the desired property. $\rfloor$

Now we are going to establish a relation between the notions of a u-measurable set and of a Polish space. The fact that the functions V_n of our DM's need not be $\mathfrak{H}_n$-measurable is intimately connected with the fact discovered by Souslin in 1917 that the continuous image of a Borel set in a Polish space need not be a Borel set. (An example for this fact may be found e.g. in Bourbaki (58),p.128). Fortunately, such continuous images turn out to be u-measurable. We shall state now the relevant definitions and results the proofs of which may be found e.g. in Bourbaki (58).

Definition. A metric space is called a *Souslin-space* if it is the continuous image of a Polish space. A subset F of a metric space is called a *Souslin set* or *analytic set*, if F, regarded as subspace, is a Souslin space.

Remarks. (i) Any Polish space is a Souslin space and the continuous image of a Souslin space into a metric space is a Souslin space.
(ii) The topological product of a countable family of Souslin spaces is a Souslin space.

The main tool for proving the universal measurability of V_n in the last coordinate is

Theorem 12.8 (cf.e.g.Bourbaki (58) p.127).
In a Souslin space, any Borel set is analytic and any analytic set is u-measurable.

Corollary 12.9. The continuous image of a Borel set in a Souslin space into a metric space is analytic, hence u-measurable.

Remark. The complement of an analytic set need not be analytic but it is u-measurable.

We shall need the following result, stated without proof in Dubins/Freedman (64),p.1214.

Theorem 12.10. Let E be a compact metric space, let M be the set of finite measures on the σ-algebra of Borel sets in E. Then the $*$-σ-algebra in M is generated by a Polish topology.

Definition. Let $(\Phi,\mathcal{F})$ and $(\Gamma,\mathcal{G})$ be two measurable spaces. A map $X:\Phi\to\Gamma$ is called an *$\mathcal{F}$-$\mathcal{G}$-Borel-isomorphism* if X is a bijection and if X and its inverse are measurable.

Remarks. (i) Let $(\Phi,\mathcal{T}),(\Gamma,\mathcal{W})$ be two topological spaces and $X:\Phi\to\Gamma$ be a homeomorphism. Then X is a $\sigma(\mathcal{T})$-$\sigma(\mathcal{W})$-Borel-isomorphism. The converse is not true.
(ii) Let $(\Phi,\mathcal{T})$ be a topological space and $(\Gamma,\mathcal{G})$ be a measurable space. If $X:\Phi\to\Gamma$ is a $\sigma(\mathcal{T})$-$\mathcal{G}$-Borel-isomorphism, then $\mathcal{G}$ is generated by the topology $X(\mathcal{T})$.
(iii) If $X:\Phi\to\Gamma$ is an $\mathcal{F}$-$\mathcal{G}$-Borel-isomorphism, then $X^{-1}(\mathcal{G})=\mathcal{F}$.

The following theorem is due to D. Rhenius who will publish a proof of it elsewhere.

Theorem 12.11. The Borel-isomorphic image of a standard-Borel-space is a standard Borel space.

Remarks. (i) Mackey (57) calls a measurable space a standard Borel space if it is the Borel-isomorphic image of a measurable space which is a SB-space in the sense of our definition. Theorem 12.8 shows that our notation is consistent with that of Mackey.
(ii) Theorem 12.11 is not absolutely necessary for the development of the theory, but it simplifies the proofs of some theorems.
(iii) The measure-theoretic product of a countable family of SB-spaces is an SB-space (cf.Mackey (57),p.138).

We shall use without proof the following lemma which goes back to J.von Neumann.

Lemma 12.12 (Mackey (57), theorem 6.3).

Let $(X_i,\mathcal{F}_i)$ i=1,2, be SB-spaces, let μ be a probability on $\mathcal{F}_1$, let B be a set in $\mathcal{F}_1 \otimes \mathcal{F}_2$. If $B_x \neq \emptyset$ $\forall x \in X_1$, then there exists a set $N \in \mathcal{F}_1$ of μ-measure zero and a measurable map $\gamma = X_1 \to X_2$ such that $\gamma(x) \in S_x$ $\forall x \in X_1 - N$.

The following theorem 12.13 has been stated by Strauch (66), p.884; for the proof only reference to the theorem of Dubins/Freedman (theorem 12.10 above) in which compactness of E is assumed, has been made. We shall give a complete proof below.

Theorem 12.13. Let W be the set of probability measures on the σ-algebra of a standard Borel space, let $\mathcal{W}$ be the *-σ-algebra in W. Then $(W,\mathcal{W})$ is a standard Borel space.

Proof. Denote by $(\Phi,\mathcal{F})$ the SB-space for which W is the set of probability measures on $\mathcal{F}$. Let us first regard the special case where $\mathcal{F}$ is generated by a Polish topology (cf.fig.4).

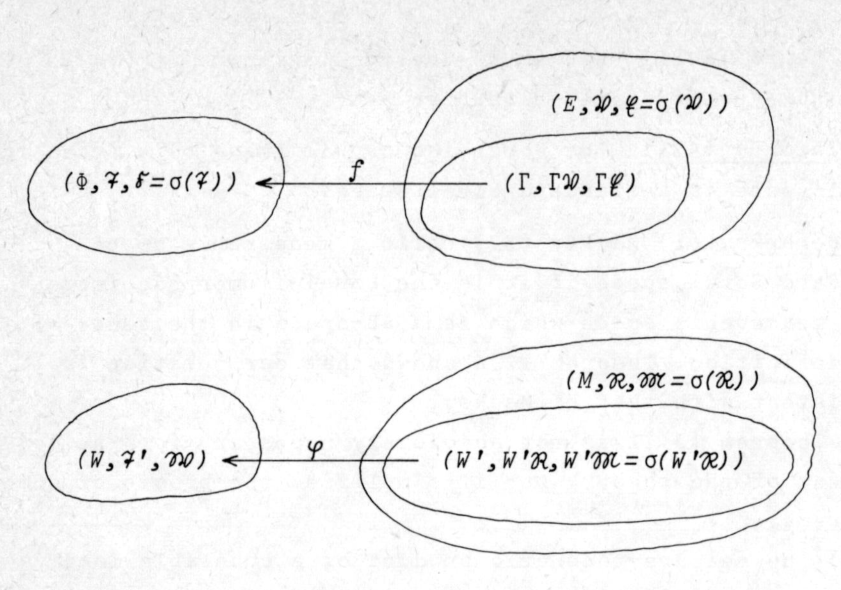

fig. 4

a) Let E be the cartesian product $\langle 0,1\rangle^{\mathbb{N}}$ and $\mathcal{W}$ the usual product topology in E. According to a wellknown theorem (cf.e.g.Bourbaki (58),p.124) there exists a set Γ in E which is the intersection of a countable family of open sets such that $(\Phi,\mathcal{T})$ and $(\Gamma,\Gamma\cap\mathcal{W})$ are homeomorphic. Let $f:\Gamma\to\Phi$ be such a homeomorphism. According to Tychonoff's theorem we know that $(E,\mathcal{W})$ is compact. Let M be the set of finite measures on $\mathcal{E}:=\sigma(\mathcal{W})$. According to a theorem of Dubins and Freedman (theorem 12.10 above) the $*$-σ-algebra $\mathfrak{M}$ in M is generated by a Polish topology $\mathcal{R}$. In particular, $(M,\mathfrak{M})$ is an SB-space. Define $W':=\{\mu\in M:\mu(\Gamma)=1,\mu(X-\Gamma)=0\}$. Obviously $W'\in\mathfrak{M}$, hence $(W',W'\mathfrak{M})$ is an SB-space.

b) Now we define a map $\varphi:W'\to W$ by means of the homeomorphism f in the following way:

(12.4) $\quad \varphi(\mu)$:= image under f of the restriction of μ on $\Gamma\mathcal{E}=\sigma(\Gamma\mathcal{W})$.

In other words, $\varphi(\mu)$ is that probability measure on $\mathcal{F}$ which assumes at $F\in\mathcal{F}$ the value $\varphi(\mu)(F)=\mu(f^{-1}(F))$. Now we shall prove that φ is a $W'\mathfrak{M}$-$\mathfrak{W}$-Borel-isomorphism, from which theorem 12.13 follows by means of theorem 12.11.

b_1) φ is surjective: Denote the (measurable) inverse of f by $\tilde{f}$. Let w be an arbitrary element of W. Then $\mu(B):=w(\tilde{f}^{-1}(B))=$ $=w(f(B)),B\in\mathcal{E}$, defines an element μ of W' with the property that $\varphi(\mu)(F)=w(F), F\in\mathcal{F}$, hence $\varphi(\mu)=w$.

b_2) φ is injective: If μ and μ' are elements of W' such that $\varphi(\mu)=\varphi(\mu')$, then $\mu(f^{-1}(F))=\mu'(f^{-1}(F)), F\in\mathcal{F}$. Since $f^{-1}(\mathcal{F})=\Gamma\mathcal{E}$, we have $\mu=\mu'$ on $\Gamma\mathcal{E}$, which implies $\mu=\mu'$.

b_3) φ is $W'\mathfrak{M}$-$\mathfrak{W}$-measurable: $\mathfrak{W}$ is generated by the system $\mathcal{D}:=(\{w\in W:w(F)\in B\}, F\in\mathcal{F}, B\in\mathcal{L}_1)$. Now we have $\varphi^{-1}(\{w\in W:w(F)\in B\})=\{\mu\in W':\varphi(\mu)(F)\in B\}=\{\mu\in W':\mu(f^{-1}(F))\in B\}\in W'\mathfrak{M}$, hence $\varphi^{-1}(\mathcal{D})\subset W'\mathfrak{M}$.

b_4) As in b_3) one shows easily that the inverse $\tilde{\varphi}$ of φ is $\mathfrak{W}$-$W'\mathfrak{M}$-measurable. Hence the proof of theorem 12.13 is complete for the case where $\mathcal{F}$ is generated by a Polish topology $\mathcal{T}$.

Let us now consider the general case. Let $(\Phi,\mathcal{F})$ and $(W,\mathfrak{W})$ be defined as above. Let G be a Polish space with topology $\mathcal{G}$ such that $\Phi\in\sigma(\mathcal{G})$ and $\mathcal{F}=\Phi\cap\sigma(\mathcal{G})$. Let W_1 be the set of probabilities on $\sigma(\mathcal{G})$ with the $*$-σ-algebra $\mathfrak{W}_1$. From the first part of the proof we know that $(W_1,\mathfrak{W}_1)$ is an SB-space. The set $W_2:=\{w\in W_1:w(\Phi)=1\}$ belongs to $\mathfrak{W}_1$, hence $(W_2,W_2\mathfrak{W}_1)$ is an SB-space. Finally the map $w\to X(w)$ from W into W_2, defined by $X(w)(B):=w(B\cap\Phi), B\in\sigma(\mathcal{G})$, is a $\mathfrak{W}$-$W_2\mathfrak{W}_1$-Borel-isomorphism. It follows from theorem 12.11 that $(W,\mathfrak{W})$ is an SB-space. $\lrcorner$

13. Universal measurability of the maximal conditional expected reward.

Definition. The DM $((S,\mathcal{S}),(A,\mathfrak{A}),D,(q_n),(r_n))$ is called a *standard decision model* if the measurable spaces $(S,\mathcal{S})$ and $(A,\mathfrak{A})$ are standard Borel spaces.

From now on we shall make the following

General assumption. All decision models considered in this and the following sections are assumed to be standard models, unless the contrary is stated.

For any $t\in\mathbb{N}, h\in\overline{H}_t, \pi\in\Delta^r$ there is defined a probability measure $Q_{t\pi}(h)$ on $\mathcal{G}:=\mathfrak{A}\otimes\mathcal{S}\otimes\mathfrak{A}\otimes\ldots$ by means of

$$(13.1) \qquad Q_{t\pi}(h) := (\pi_t q_t \pi_{t+1}\ldots)(h,\cdot) .$$

It follows from lemma A9 that $(h,B)\to Q_{t\pi}(h)(B)$, $t\in\overline{H}_n, B\in\mathcal{G}$, is a conditional P_π-distribution of $(\alpha_t,\zeta_{t+1},\alpha_{t+1},\ldots)$ under the condition χ_t. $Q_{t\pi}$ is a map from $\overline{H}_t$ to the set W of probability measures on $\mathcal{G}$.

Lemma 13.1. (cf. Strauch (66) lemma 7.2).
Let W be the set of probability measures on $\mathfrak{A}\otimes\mathcal{S}\otimes\mathfrak{A}\otimes\ldots$ with the $*$-σ-algebra $\mathfrak{W}$. Then $\bigcup_{\pi\in\Delta^r}$ graph $Q_{t\pi}$ belongs to $\tilde{\mathcal{H}}_t \otimes \mathfrak{W}$, $t\in\mathbb{N}$.

Proof. a) Put $\Gamma:=\bigcup_\pi$ graph $Q_{t\pi}$, and $Q_\pi:=Q_{t\pi}$. The problem will be to find a characterization of Γ by means of a countable family of $\mathcal{S}\otimes\mathfrak{W}$-measurable maps (that do not depend on π). Put $F:=A_t\times S_t\times A_{t+1}\ldots$, where $A_n:=A$ and $S_n:=S$ for all $n\geq t$, and let α_n,ζ_n be the projection from F into A_n and S_n, respectively. (There is no danger of misunderstanding though α_n and ζ_n have a slightly different meaning in connection with the probability space $(\overline{H},\overline{\mathcal{H}},P_\pi)$.) As usual $w_{(\alpha_t,\ldots,\zeta_{n+1})}$ will denote the distribution of the random variable $(\alpha_t,\ldots,\zeta_{n+1})$ under the probability $w\in W$. Put $\Psi_n:=(\alpha_t,\zeta_{t+1},\ldots,\zeta_n)$, $n>t$.

b) We are going to prove the following assertion:
The point (h,w) belongs to Γ iff it satisfies

$$(13.2) \qquad w_{\Psi_{n+1}}((H_{n+1})_h) = 1 \ , \ n\geq t,$$

and

$$(13.3) \qquad w_{\Psi_{n+1}} = w_{(\Psi_n,\alpha_n)} \otimes q_n(h,\cdot), \ \ n\geq t.$$

It follows from the definition of Γ that $(h,w)\in\Gamma$ iff $w=Q_\pi(h)$ for some randomized plan π or equivalently, iff

(13.4) $\qquad w\psi_{n+1} = (\pi_t q_t \cdots q_n)(h,\cdot),\ n\geq t,$ for some $\pi\in\Delta^r$.

Condition (13.4) implies (13.2), as shown in the proof of lemma 9.1.b). It follows from (13.4) that $w_{(\Psi_n,\alpha_n)} = (\pi_t q_t \cdots \pi_n)(h,\cdot)$, hence $w_{(\Psi_n,\alpha_n)} \otimes q_n(h,\cdot)=(\pi_t q_t \cdots q_n)(h,\cdot)=w\psi_{n+1}$, hence also (13.3) is a consequence of (13.4). Now we shall assume that (13.2) and (13.3) hold, and we shall construct a $\pi\in\Delta^r$ satisfying (13.4). According to corollary 12.5 we may represent w in the form $w=u_t v_t u_{t+1} v_{t+1} \cdots$ where $u_t=w\alpha_t$, $v_n\in w_{\zeta_{n+1}|(\Psi_n,\alpha_n)}$ and $u_n\in w_{\alpha_n|\Psi_n}$, $n>t$. Corollary 12.7 tells us that we can select u_n in such a way that $u_t(D_t(h))=1$ and

(13.5) $\qquad u_n(x,D_n(h,x)) = 1 \qquad n>t, x\in(H_n)_h$.

In fact, let us use corollary 12.7 by putting $X:=A\times S\times\ldots\times S$ (2n-2t factors), $Y:=A$, $B:=\{(x,a)\in X\times Y: x\in(H_n)_h, a\in D_n(h,x)\}$, $\mu:=w_{(\Psi_n,\alpha_n)}$, $q:=\sigma_n(h,\cdot)$, where σ is an arbitrary randomized plan. Then we have $\mu(B)=w\psi_{n+1}((H_{n+1})_h)=1$ by (13.2) and $q(x,B_x)=\sigma_n(h,x,D_n(h,x))=1$ for $x\in pr_1(B)=(H_n)_h$, and (13.5) follows from corollary 12.7.

Now we define the randomized plan π by $\pi_n:=\sigma_n$, $n<t$,

$$\pi_n(h_t,x,\cdot):=\begin{cases} u_n(x,\cdot) & \text{for } h_t=h, \\ \sigma_n(h_t,x,\cdot) & \text{otherwise} \end{cases} \quad n\geq t$$

Since $\{h\}$ belongs to $\bar{\mathfrak{F}}_t$ and since (13.5) holds, π belongs to Δ^r. We have to verify that π satisfies (13.4). This is true for $n=t$, as (13.3) implies $w\psi_{t+1}=w_{(\alpha_t,s_{t+1})}=w\alpha_t q_t(h,\cdot)= u_t q_t(h,\cdot)=\pi_t q_t(h,\cdot)$. If (13.4) is assumed to be true for some $n\geq t$, then again (13.3) implies

$$w\psi_{n+1} = w_{(\Psi_n,\alpha_n)} q_n(h,\cdot) = w\psi_n u_n q_n(h,\cdot) = $$
$$= w\psi_n \pi_n q_n(h,\cdot) = (\pi_t q_t \cdots \pi_n q_n)(h,\cdot).$$

Hence we have shown that (h,w) belongs to Γ iff (13.2) and (13.3) is fulfilled.

c) It is well-known (cf.e.g.Bauer (68),p.258) that the algebra generated by a countable family of sets is also countable.

It follows easily that the σ-algebra of Borel sets in a topological space with countable base is generated by a countable algebra. The σ-algebra $\mathcal{B}$ of Borel sets in an SB-space E is generated by the trace on E of a Polish topology, hence $\mathcal{B}$ is generated by a countable algebra. Let now $\mathcal{L}_n$ be a countable algebra that generates $\mathcal{G}_n := \mathcal{A}\otimes\mathcal{S}\otimes\ldots\otimes\mathcal{S}$ (2n-2t factors). Since two probabilities on $\mathcal{G}_n$ which coincide on $\mathcal{L}_n$, are identical, we know that (13.3) holds whenever (13.3) holds on $\mathcal{L}_{n+1}$. It is now obvious from (13.2) and (13.3) that Γ belongs to $\overline{\mathcal{H}}_t\otimes\mathcal{W}$ if we can show that for any $n\in\mathbb{N}$ and any $C\in\mathcal{L}_{n+1}$ the following three real functions are $\overline{\mathcal{H}}_t\otimes\mathcal{W}$-measurable:

(13.6) $\qquad (h,w) \to w_{\Psi_{n+1}}((H_{n+1})_h)$,

(13.7) $\qquad (h,w) \to w_{\Psi_{n+1}}(C)$,

(13.8) $\qquad (h,w) \to w_{(\Psi_n,\alpha_n)}q_n(h,C)$.

Now we have $w_{\Psi_{n+1}}((H_{n+1})_h) = \int 1_{H_{n+1}}(h,y)w_{\Psi_{n+1}}(dy)$. The measurability of the map (13.6) follows then from corollary 12.3 since $(h,y)\to 1_{H_{n+1}}(h,y)$ is measurable. The measurability of the maps (13.7) and (13.8) follows in a similar way.

Now we can prove our main result of this section.

Theorem 13.2. (cp.Strauch (66)). Let $V_n:H_n\to\overline{\mathbb{R}}$ be the maximal conditional expected reward for time period (t,∞) within the set of randomized plans. Then V_n is $\overline{\mathcal{H}}_n$-universally measurable and $V_{n+1}(h,a,\cdot)$ is $\mathcal{S}$-universally measurable for $n\in\mathbb{N}$, $(h,a)\in K_n$.

Proof. a) Put $X:=A\times S\times A\ldots$ and define $(W,\mathcal{W})$ and Γ as in lemma 13.1. According to corollary 12.3 the map $v:\overline{H}_n\times W\to\overline{\mathbb{R}}$ defined by

$$v(h,w):=\int\Big(\sum_n^\infty r_i\circ(\chi_i,\alpha_i)\Big)(h,x)w(dx),\ (h,w)\in\overline{H}_n\times W,$$

is $\overline{\mathcal{H}}_n\otimes\mathcal{W}$-measurable. Therefore lemma 13.1 implies that the set $B_\alpha:=\Gamma\cap\{(h,w)\in\overline{H}_n\times W: v(h,w)>\alpha\}$ belongs to $\mathcal{S}\otimes\mathcal{W}$ for any real α. $(\overline{H}_n,\overline{\mathcal{H}}_n)$ is an SB-space by remark (iii) after theorem 12.11. $(W,\mathcal{W})$ is an SB-space by theorem 12.13. Hence (cf.again remark (iii) after theorem 12.11) $(\overline{H}_n\times W,\overline{\mathcal{H}}_n\otimes\mathcal{W})$ is an SB-space, which means that $\overline{H}_n\times W$ is a Borel set in a Polish space hence in some Souslin space. Theorem 12.8 tells us that therefore $\overline{H}_n \times W$ is also a Souslin space. Let pr_1 be the projection from $\overline{H}_n \times W$ into $\overline{H}_n$ which is a continuous map. Corollary 12.9 implies, since $B_\alpha \in \overline{H}_n \otimes \mathcal{W}$, that $pr_1(B_\alpha)$ is measurable with respect to the σ-algebra $\hat{\mathcal{H}}_n$ of $\overline{\mathcal{H}}_n$-universally measurable sets. The $\hat{\mathcal{H}}_n$-measurability of

$$pr_1(B_\alpha) = \{h \in \overline{H}_n: \ V_{n\pi}(h) > \alpha \text{ for some } \pi \in \Delta^r\}$$

$$= \{h \in \overline{H}_n: \ V_n(h) > \alpha\}, \alpha \in \mathbb{R},$$

implies the $\hat{\mathcal{H}}_n$-measurability of V_n.

b) One will be tempted to consider the $\mathcal{Y}$-universal measurability of $V_{n+1}(h,a,\cdot)$ as trivial, since the latter is a section of the $\overline{\mathcal{H}}_{n+1}$-universally measurable function. However, this argument, valid for the product of measurable spaces, fails in the case of universal measurability.

We shall use the standard reduction method, developped in section 3, for the reduction of the statement for $V_{n+1}(h,a,\cdot)$ to that for V_1. Let us fix $t \in \mathbb{N}, k \in K_t$, and let us define a new DM$((S,\mathcal{Y}),(A,\mathcal{O}),D',(q_n'),(r_n'))$ in exactly the same way as in section 3. Let us associate with any randomized plan $\pi \in \Delta^r$ a sequence $\pi' = (\pi_n')$ by $\pi_n'(h',\cdot) := \pi_{n+1}(k,n',\cdot)$, $n \in \mathbb{N}, h' \in H_n'$. Then $\pi \to \pi'$ is a map from Δ^r onto $\Delta^{r'}$ such that $V_{n+1,\pi}(k,s) = V_{1\pi'}'(s)$, $s \in S$. Therefore $V_{n+1}(k,s) = \sup_{\pi \in \Delta^r} V_{n+1,\pi}(k,s) = \sup_{\pi' \in \Delta^{r'}} V_{1\pi'}'(s) = V_1'(s)$. Hence the u-measurability of $V_{n+1}(k,\cdot)$ follows from that of V_1'. ⌋

14. The optimality equation.

Denote by N_n^+ and N_n^- the set of functions $v:H_n \to \overline{\mathbb{R}}$ that are u-measurable in the last coordinate and which are bounded from below or from above, respectively. The operator U_n, defined by

$$(14.1) \qquad U_n v := \sup_{a \in D_n(\cdot)} \left[r_n(\cdot,a) + \int q_n(\cdot,a,ds) v(\cdot,a,s) \right]$$

is defined on N_{n+1}^+ or N_{n+1}^- in case (EP) and (EN), respectively, but $U_n v$ does not necessarily belong to N_n^+ or N_n^-.

Definition. A solution of the OE is a sequence (v_n) of maps $v_n \in N_n^+$ or $v_n \in N_n^-$ in case (EP) and (EN), respectively, such that

$$v_n = U_n v_{n+1}, \; n \in \mathbb{N}.$$

Our goal will be to get results analogous to those of section 3. As a preparation we prove

Theorem 14.1 (cp. Strauch (66), theorem 8.1). If $p(\{s:V_1(s)=\infty\})=0$, then there exists for any $\varepsilon>0$ a randomized plan σ such that

$$(14.2) \qquad p(\{s \in S: \; V_{1\sigma}(s) \geq V_1(s)\}-\varepsilon) = 1.$$

Remark. A plan σ satisfying (14.2) is called a (p,ε)-optimal plan by Strauch (66). A similar concept was already introduced and studied by Blackwell (65).

Proof. a) Since V_1 is u-measurable by theorem 13.2 there exists, according to lemma 12.1., a set $N \in \mathfrak{T}$ of p-measure zero and a measurable map $Z:S \to \overline{\mathbb{R}}$ such that $Z=V_1$ on N^c. We may assume without loss of generality that N contains the set $\{s \in S: Z_1(s)=\infty\}$. Let W denote the set of probabilities on $\mathfrak{A} \otimes \mathfrak{T} \otimes \mathfrak{A} \otimes \ldots$ with the *-σ-algebra $\mathfrak{W}$. We have shown in the proof of theorem 13.2 that the map

$$(s,w) \to v(s,w) := \int \left(\sum_1^\infty r_i \circ (\chi_i, \alpha_i)\right)(s,x) w(dx)$$

is $\mathfrak{T} \otimes \mathfrak{W}$-measurable. Now we are going to use the set Γ, defined in the proof of lemma 13.1. for arbitrary $t \in \mathbb{N}$, for $t=1$. Define the set

$$\Gamma_\varepsilon := \Gamma \cap \{(s,w) \in S \times W: \text{"} s \in N^c \text{ and } v(s,w) \geq Z(s)-\varepsilon \text{" or "} s \in N \text{"} \}$$

which obviously belongs to $\mathfrak{T} \otimes \mathfrak{W}$. For any $s \in S$ the section

$(\Gamma_\varepsilon)_s$ is not empty. In fact, $\Gamma_s \neq \emptyset$ by the definition of Γ. Moreover, if $s \in N^c$ then $V_1(s)=Z(s)<\infty$ and there exists by the definition of V_1 a plan $\pi \in \Delta^r$ (depending on s) such that $v(s,w)=V_{1\pi}(s) \geq V_1(s)-\varepsilon=Z(s)-\varepsilon$ for $w:=Q_{1\pi}(s)$. Having verified the $(\Gamma_\varepsilon)_s \neq \emptyset$ $\forall s \in S$, we conclude from lemma 12.12 that there exists a set $N_1 \in \mathcal{T}$ of p-measure zero and a measurable map $\gamma: S \to W$ such that $\gamma(s) \in (\Gamma_\varepsilon)_s$ $\forall s \in N_1^c$. Remark (iii) after the definition of a *-σ-algebra in section 12 tells us that the map

$$\nu(s,C) := \gamma(s)(C), \quad C \in \mathcal{O}\otimes\mathcal{T}\otimes\ldots,$$

is a transition probability from S to $F:=A\times S\times\ldots$.

b) Let $\alpha_n, \zeta_n, \chi_n$ have the *usual* meaning (cf. section 11). According to corollary 12.5 there exists a factorization of the probability $w:=p\otimes\nu$ of the form $w=p\sigma_1 v_1 \sigma_2 v_2 \ldots$, where $\sigma_n \in w_{\alpha_n|\chi_n}$ and $v_n \in w_{\zeta_{n+1}|(\chi_n,\alpha_n)}$. We are going to show by means of corollary 12.7 that we can choose σ_n in such a way that $\sigma:=(\sigma_n)$ belongs to Δ^r, i.e. such that $\sigma_n(h,D_n(h))=1$, $n\in\mathbb{N}, h\in H_n$. We fix n and put $(X,\mathcal{F}):=(\overline{H}_n,\overline{\mathcal{G}}_n),(Y,\mathcal{G}):=(A,\mathcal{O})$; $B:=K_n\in\overline{\mathcal{G}}_n\otimes\mathcal{O}$, $q=\pi_n$ for an arbitrary $\pi\in\Delta^r$, $\mu:=w_{(\psi_n,\alpha_n)}$. Then $q(x,B_x)=\pi_n(x,D_n(x))>0$ for all $x\in pr_1(B)=H_n$. Moreover μ is concentrated on B: for any fixed $s\in N_1^c$ we have $\gamma(s)\in\Gamma_s$, i.e. $\nu(s,\cdot)=Q_{1\pi}(s,\cdot)$ for some $\pi\in\Delta^r$ (depending on s). For $\hat{B}:=B\times S\times A\times\ldots$ we get $\nu(s,\hat{B}_s)=Q_{1\pi}(s,\hat{B}_s)=\int(\pi_1\ldots q_{n-1})(s,dx)\pi_n(s,x,D_n(s,x))=1$, hence $\mu(B)=\int p(ds)\, 1_{N_1^c}(s)\nu(s,\hat{B}_s)=1$.

c) Now we shall show that there exists a set $N_2\in\mathcal{T}$ of p-measure zero such that $\nu(s,\cdot)=Q_{1\sigma}(s)$ for $s\in N_2^c$, where $\sigma\in\Delta^r$ is the plan constructed in part b). At first, it is well-known (cf. Bauer (68), p.257) that $p\otimes\nu=p\sigma_1 v_1\sigma_2\ldots$ implies the existence of a set $N_3\in\mathcal{T}$ of p-measure zero such that $\nu(s,\cdot)=(\sigma_1 v_1\ldots)(s,\cdot)$ for $s\in N_3^c$. Put $N_2:=N_1\cup N_3$ and fix some $s\in N_2^c$. Then $\nu(s,\cdot)=\gamma(s)\in\Gamma_s$, i.e. there exists a plan $\delta\in\Delta^r$ (depending on s) such that $\nu(s,\cdot)=Q_{1\delta}(s,\cdot)=$ $=(\delta_1 q_1\delta_2\ldots)(s,\cdot)$. Lemma A8 implies, since $\nu(s,\cdot)$ equals also $(\sigma_1 v_1\sigma_2\ldots)(s,\cdot)$, that $\nu(s,\cdot)$ equals also $Q_{1\sigma}(s,\cdot)$.

d) Finally we fix some $s\in(N\cup N_2)^c$. Then $w:=Q_{1\sigma}(s)=\gamma(s)\in(\Gamma_\varepsilon)_s$ and $s\in N^c$, hence $V_{1\sigma}(s)=v(s,w)\geq Z(s)-\varepsilon=V_1(s)-\varepsilon$. _|

Theorem 14.2. Let $V:=\sup_{\pi\in\Delta^r} G_\pi$ be the maximal expected reward within the set of randomized plans. Then

$$V = \int V_1 dp.$$

Remark. The model of Strauch (66) contains the assumption that V_1 is bounded from above by a constant. In this case theorem 14.2 is an immediate consequence of theorem 14.1. We do not make such an assumption, therefore theorem 14.2, which will also be used in the proof of the OE, needs a separate proof.

Proof. a) According to lemma 11.2 we have $G_\pi=\int V_{1\pi}dp\leq$ $\leq\int V_1 dp$, hence $V\leq\int V_1 dp$.

b) Let us assume that $p(\{s\in S:V_1(s)=\infty\})=0$. It follows from theorem 14.1 that there exists for any $\varepsilon>0$ a $\sigma\in\Delta^r$ such that $V\geq G_\pi=\int V_{1\pi}dp\geq\int V_1 dp-\varepsilon$, hence $V\geq\int V_1 dp$.

c) Let us assume that $p(\{s\in S:V_1(s)=\infty\})>0$, hence $\int V_1 dp=\infty$. Now we shall use a modification of the proof of theorem 14.1. Let $W,\mathfrak{W},N,Z,v$ be defined as in the proof of theorem 14.1. The set $M:=\{s\in S:Z(s)=\infty\}$ and hence also the set MN^c belongs to $\mathfrak{T}$ and has positive p-measure. Let us define for fixed $m\in\mathbb{N}$ the set $\Gamma_m:=\Gamma\cap\{(s,w)\in S\times W$: "$s\in MN^c$ and $v(s,w)\geq m$" or "$s\notin MN^c$"$\}$. Γ_m belongs to $\mathfrak{T}\otimes\mathfrak{W}$, and for any $s\in S$ the section $(\Gamma_m)_s$ is not empty. In fact, if $s\in MN^c$ then there exists - since $V_1(s)=\infty$ - a plan $\pi\in\Delta^r$ (depending on s) such that $V_{1\pi}(s)\geq m$, hence $w:=Q_{1\pi}(s)$ satisfies $v(s,w)=V_{1\pi}(s)\geq m$. It follows as in the proof of theorem 14.1 that there exists a set $N_4\in\mathfrak{T}$ of p-measure zero and a plan $\sigma\in\Delta^r$ such that $Q_{1\sigma}(s)\in(\Gamma_m)_s$ for $s\in N_4^c$, i.e. $Q_{1\sigma}(s)\geq m$ for $s\in MN_4^c$, which is a set of positive p-measure. Therefore $V\geq G_\sigma\geq m\cdot p(M)$. The preceding reasoning is valid for any $m\in\mathbb{N}$, hence $V=\infty$. _|

The following lemma is obtained from parts b) and c) of the proof of theorem 14.2 by the standard reduction method.

Lemma 14.3. For any $\varepsilon>0$, $m,n\in\mathbb{N}$, $(h,a)\in K_n$ there exists a randomized plan τ such that

$$(14.3)\quad \int q_n(h,a,ds)V_{n+1,\tau}(h,a,s)\geq\min(m,\int q_n(h,a,ds)V_{n+1}(h,a,s))-\varepsilon.$$

Theorem 14.4. (cp. Strauch (66) The sequence (V_n) satisfies the OE

$$(14.4) \qquad V_n := \sup_{a \in D_n(\cdot)} [r_n(\cdot,a) + \int q_n(\cdot,a,ds) V_{n+1}(\cdot,a,s)] =: U_n V_{n+1}, \quad n \in \mathbb{N}.$$

Proof. a) As in the proof of theorem 3.9 one shows easily with the aid of the formula $V_{n\pi} = \Lambda_{n\pi} V_{n+1,\pi}$, that $V_n \leq U_n V_{n+1}$.

b) We fix $h \in H, b \in D_n(h)$. According to lemma 14.3 there exists a $\tau \in \Delta^r$ (depending on (h,b)), such that $\int q_n(h,a,ds) V_{n+1,\tau}(h,b,ds) \geq \min(m, \int q_n(h,b,ds) V_{n+1}(h,b,s)) - \varepsilon$. Put

$$\sigma_n(h_n,B) := \begin{cases} 1_B(b) & \text{if } h_n = h \\ \tau_n(h_n,B), & \text{otherwise,} \end{cases} \qquad B \in \mathfrak{A}.$$

$\sigma_n(\cdot,B)$ is measurable since $\mathfrak{A}$ contains all singletons. Therefore $\sigma := (\tau_1, \tau_2, \ldots, \tau_{n-1}, \sigma_n, \tau_{n+1}, \ldots)$ belongs to Δ^r, and

$$V_n(h) \geq V_{n\sigma}(h) = \int \sigma_n(h,da) [r_n(h,a) + \int q_n(h,a,ds) V_{n+1,\sigma}(h,a,s)]$$
$$\geq r_n(h,b) + \min(m, \int q_n(h,b,ds) V_{n+1}(h,b,s)) - \varepsilon.$$

This inequality being true for any $m \in \mathbb{N}$ and any $\varepsilon > 0$, we have $V_n \geq U_n V_{n+1}$. $\rfloor$

Theorem 4.1 and its proof carry over with only slight changes and yield

Theorem 14.5. There is defined the double sequence $(V_{nk}, n \in \mathbb{N}, k \in \mathbb{N})$ of functions $V_{nk} \in N_n^+$ by

$$V_{no} :\equiv 0,$$
$$V_{nk} := U_n V_{n+1,k-1}, \quad n \in \mathbb{N}, k \in \mathbb{N}.$$

(V_{nk}) has the property

$$(14.5) \qquad V_{nk}(h) = \sup_{\pi \in \Delta^r} E_\pi [\sum_{i=n}^{n+k-1} r_i \circ (\chi_i, \alpha_i) | \chi_n = h], \quad n \in \mathbb{N}, k \in \mathbb{N}, h \in H_n,$$

and, in case (EP), we have

$$(14.6) \qquad V_n = \lim_k V_{nk}, \quad n \in \mathbb{N}.$$

Now we shall reproduce the example of Strauch (66) for a DM, in which V_n is not measurable (but u-measurable). Let B be a Borel subset in $\mathbb{R}^2$, which is contained in $(0,1)^2$ and whose projection D into $\mathbb{R}$ is not a Borel set. (The existence of such a set is proved, e.g. in Hausdorff (27) p.177.) B as well as $Z:=\langle 0,1)$ are SB-spaces when endowed with their σ-algebra of Borel sets. Hence $S:=B+Z$ is also an SB-space (Mackey (57),p.138). Now we consider the DM $(S,A,D,(q_n),r_n)$ where: S as above; $A:=(0,1)$; $D_n(h):=A$; q_o arbitrary; $q_n(h,a,\cdot):=q(s_n,a,\cdot)$ where $q((x,y),a,\cdot)=\delta(x)$, $(x,y)\in B$, and $q(z,a,\cdot)=\delta(0)$, $z\in Z$; $r_n(h,a)=\beta^{n-1}1_B(s_n,a)$ for some $\beta\in(0,1\rangle$. A straightforward computation yields

$V_{n1}(h) = U_n 0(h)=\beta^{n-1}1_D(s_n)$,

$V_{n2}(h) = U_n V_{n+1,1}(h) = \beta^{n-1}1_D(s_n)+\beta^n 1_B(s_n)$,

$V_{n3} = V_{n2}$,

hence $V_n(h)=V_{n2}(h)$. Since D is not a Borel set in $\mathbb{R}$, it does not belong - considered as a subset of S - to the direct product of the σ-algebras of Borel sets in B and Z.

Also theorem 3.11, its proof and the remarks following it carry over to

Theorem 14.6. (i) In case (EP) the sequence (V_n) is the (termwise) smallest of those solutions (v_n) of the OE that satisfy

$$\lim_n \|v_n^-\| = 0. \tag{14.7}$$

(ii) In case (C) the sequence (V_n) is the only solution (v_n) of the OE that satisfies

$$\lim_n \|v_n\| = 0. \tag{14.8}$$

15. Substitution of randomized plans by deterministic plans.

In section 9 we have shown that if the state space S and the action space A are countable, then $V_n=G_n$ and $V=G$. Moreover, in case (EN) there exists for any randomized plan a deterministic plan g such that $G_g \geq G$.

In this section we shall investigate the same problems under the assumption that $(S,\mathcal{T})$ and $(A,\mathcal{O})$ are SB-spaces.

We begin with a slight extension of a result of Blackwell (64).

Lemma 15.1. Let $(X,\mathcal{F})$ and $(Y,\mathcal{G})$ be SB-spaces, let B be a set in $\mathcal{F}\otimes\mathcal{G}$ and let q be a transition probability from X to Y such that $q(x,B_x)=1 \quad \forall x\in X$. Let $v:X\times Y\to \overline{\mathbb{R}}$ be measurable and $\int q(x,dy)v^+(x,y)<\infty$, $x\in X$. Then there exists a measurable map $f:X\to Y$ such that

$$(15.1) \qquad f(x)\in B_x, \quad x\in X.$$

$$(15.2) \qquad v(x,f(x)) \geq \int q(x,dy)v(x,y), \quad x\in X.$$

Proof. (cp.Blackwell (64)). The map

$$(15.3) \qquad u := v1_B + (-\infty)1_{B^c}$$

is measurable, and $\int q(x,dy)u^{\pm}(x,y)=\int_{B_x} q(x,y)u^{\pm}(x,y)=\int q(x,y)v^{\pm}(x,y)$. In particular, we have $\int q(x,dy)u(x,y)=\int q(x,dy)v(x,y)=:W(x)$, $x\in X$. W is measurable, hence $F:=\{(x,y)\in X\times Y:u(x,y)\geq W(x)\}$ belongs to $\mathcal{F}\otimes\mathcal{G}$. We have $q(x,F_x)>0$ for all $x\in X$. In fact, if $q(x,F_x)=0$ for some fixed x, then $u(x,\cdot)<\int q(x,dz)u(x,z)\leq\int q(x,dz)u^+(x,z)<\infty$ holds for a set of points y of $q(x,\cdot)$-measure one, hence $\int q(x,dy)u(x,y)\leq\int q(x,dz)u(x,z)-\varepsilon$ for some positive ε, which is a contradiction. We have also $q(x,B_x)>0$, $x\in X$, from our assumptions. According to lemma 12.6 there exist measurable maps $g:X\to Y$ and $\varphi:X\to Y$ such that $\varphi(x)\in B_x$ and $g(x)\in F_x$, i.e.

$$(15.4) \qquad u(x,g(x)) \geq W(x), \ x\in X.$$

Now we shall show that

$$f := g\cdot 1_{[W>-\infty]} + \varphi\cdot 1_{[W=-\infty]}$$

has the desired properties (15.1) and (15.2). Take an arbitrary $x\in X$. If $W(x)=-\infty$, then $f(x)=\varphi(x)\in B_x$, and (15.2) is trivially true. Now let us assume that $W(x)>-\infty$, hence

$f(x)=g(x)$. It follows from (15.4) that $u(x,g(x))>-\infty$, hence $g(x)\in B_x$, according to (15.3). Finally, using (15.3) and (15.4) again, we get $v(x,f(x))=v(x,g(x))=u(x,g(x))\geq W(x)$. ⌋

<u>Theorem 15.2.</u>(cp.Strauch (66)). In case (EN) there exists for any randomized plan σ a deterministic plan $f\in\Delta$ such that $G_f\geq G_\sigma$ and $G_{nf}\geq V_{n\sigma}\circ h_{nf}$, $n\in\mathbb{N}$.

The proof is nothing but a copy of the proof of theorem 9.4 and lemma 9.3, whereby lemma 15.1 is used in the following way: For fixed $n\in\mathbb{N}$ we put $(X,\mathfrak{f}):=(\overline{H}_{\infty n},\overline{\mathfrak{G}}_n)$, $(Y,\mathfrak{g}):=(A,\mathfrak{A})$, $B:=K_n$, $q(h,\cdot):=\sigma_n(h,\cdot)$, $v(h,a):=E_\sigma[\sum_n r_\nu\circ(\chi_\nu,\alpha_r)|(\chi_n,\alpha_n)=(h,a)]$. Then the map f of lemma 15.1 plays the role of the map g_n in the proof of theorem 9.4. ⌋

<u>Lemma 15.3</u>. The map $g\rightarrow f$ from the set Δ' of H-plans into the set Δ of (deterministic) plans, defined by

$$f_n := g_n\circ h_{nf}\ ,\quad n\in\mathbb{N},$$

is surjective.

Proof. Let f be an arbitrary plan in Δ, let g' be an arbitrary H-plan and put for $n\in\mathbb{N}$, $h\in\overline{H}_n$

$$g_n(h) := \begin{cases} f_n(y)\ , & \text{if } h\in h_{nf}(S^n), \\ g'_n(h)\ , & \text{otherwise.} \end{cases}$$

Then $g_n(h)\in D_n(h)$ and $f_n=g_n\circ h_{nf}$, $n\in\mathbb{N}$, $h\in H_n$. It remains to prove that g_n is measurable. Since $h\rightarrow f_n(y)$ and $h\rightarrow g'_n(h)$ are measurable maps from H_n into A, it is sufficient to show that $h_{nf}(S^n)$ is a measurable subset of H_n. Now we have

$$h_{nf}(S^n) = \bigcap_{\nu=1}^{n-1}\ \{h\in\overline{H}_n:\ f_\nu(y_\nu)=a_\nu\} = \bigcap_{\nu=1}^{n-1}\ S^{n-\nu}\times A^{n-2}\times \text{graph } f_\nu.$$

Therefore we have to show that graph f_ν is a measurable subset of $S^\nu\times A$. This follows immediately from a well-known theorem (cf.e.g. Bourbaki (58),p.148), since $\bigotimes_1^\nu \mathfrak{T}$ and $\mathfrak{A}$ are generated by metrisable topologies which have a countable base. ⌋

It follows from the preceding lemma that

$$G = \sup_{f\in\Delta} G_f = \sup_{g\in\Delta'} G_g \leq V$$

and

$$G_n(h) = \sup_{f\in\Delta_n(h)} G_{nf}(y) = \sup_{g\in\Delta'} G_{ng}(h) \leq V_n(h).$$

On the other hand, in case (EN) we have $G_n \geq V_n$ and $G \geq V$ by theorem 15.2, hence $G=V$ and $G_n=V_n$. We conjecture that this is true also in case (EP), but we can prove it only under somewhat stronger conditions. We remark that if one could show the u-measurability of G_n, then one could carry over the proof of theorem 9.4, from which $G_n=V_n$ would follow.

Theorem 15.4. If each of the reward functions r_n is bounded from above then $G=V$ and $G_n=V_n$, $n \in \mathbb{N}$.

Proof. According to the reasoning after theorem 15.2 we have only to consider the case (EP). The reduction carried through at the end of the proof of theorem 4.1 shows that we may even assume $r_n \geq 0$. If we put

$$G_{nk}(h) := \sup_{f \in \Delta_n(h)} E_f\left[\sum_{i=n}^{n+k-1} r_{if} \circ \eta_i \mid \eta_n = y\right]$$

then we may apply theorem 15.2 in order to get

$$G_{nk}(h) = V_{nk}(h) := \sup_{\pi \in \Delta^r} E_\pi\left[\sum_{i=n}^{n+k-1} r_i \circ (\chi_i, \alpha_i) \mid \chi_n = h\right],$$

for any $n \in \mathbb{N}$, $k \in \mathbb{N}$. Lemma 3.4 and the monotone convergence theorem yield $G_n := \lim_k G_{nk}$ and $V_n := \lim_k V_{nk}$, hence $V_n = G_n$. ┘

16. A generalization of the fixed point theorem for contractions.

Blackwell (65), Maitra (68) and others use the fixed point theorem for contractions for deriving some results. We shall present in this section a generalization of the fixed point theorem that will play the same role in our general model. Our exposition follows closely that of Hinderer (67).

Let (B_n, ρ_n) be an arbitrary metric space with metric ρ_n; $B := \overset{\infty}{\underset{1}{\times}} B_n$. Let (T_n) be a sequence of maps $T_n : B_{n+1} \to B_n$, and define a map $T:B\to B$ by

$$(16.1) \qquad T(b_n) := (T_n b_{n+1}) \ , \quad (b_n)\in B.$$

Let $\beta := (\beta_{mn}, m\in \mathbb{N}, n\in\mathbb{N})$ be an infinite matrix of non-negative real numbers. For any point $c=(c_n)\in B$ we define the set

$$B(c) := \{(b_n)\in B;\ \lim_m \rho_m(b_m, c_m)=0\}.$$

Theorem 16.1. Let the metric spaces (B_m, ρ_m) be complete. Let c be an arbitrary point in B. Let us assume that the following assumptions hold:

$$(16.2) \qquad \lim_m \sup_n \beta_{mn} = 0,$$

$$(16.3) \qquad \rho_n(T_n b, T_n d) \le \rho_{n+1}(b,d) \ , \quad n\in\mathbb{N},\ b,d\in B_{n+1} \ , \text{ and}$$

$$(16.4) \qquad \rho_m(c_m, T_m T_{m+1} \cdots T_{m+n} c_{m+n+1}) \le \beta_{mn} \ , \quad m,n\in\mathbb{N}.$$

Then there exists within B(c) exactly one fixed point (v_n) of T. Moreover, (v_n) has the property that

$$(16.5) \qquad \rho_m(v_m, c_m) \le \varliminf_n \beta_{mn} \ , \quad m\in\mathbb{N} .$$

The proof is similar to the standard proof for the fixed point theorem for contractions. We put $v^o := c$ and define a sequence of points $v^k = (v_{nk}, n\in\mathbb{N})$, $k\in\mathbb{N}$, by $v^k := Tv^{k-1}$, i.e.

$$(16.6) \qquad v_{nk} := T_n v_{n+1,k-1} \ , \quad n\in\mathbb{N},\ k\in\mathbb{N}.$$

For $0<m<k$ we get

$$\begin{aligned} \rho_m(v_{nm}, v_{nk}) &= \rho_n(T_n T_{n+1} \cdots T_{n+m-1} v_{n+m,0}, T_n T_{n+1} \cdots T_{n+k-1} v_{n+k,0}) \\ &\le \rho_{n+m}(c_{n+m}, T_{n+m} T_{n+m+1} \cdots T_{n+k-1} c_{k+n}) \\ &\le \beta_{n+m,k-m-1} \le \sup_\nu \beta_{n+m,\nu} \to 0 \ (m\to\infty). \end{aligned}$$

Therefore $(v_{nk}, k \in \mathbb{N})$ is for any fixed n a Cauchy sequence; since (U_n, ρ_n) is complete, it converges to some point $v_n \in B_n$. Now we show that (v_n) satisfies (16.6) and belongs to B(c). From the continuity of the metric ρ_n and the inequality

$$\rho_n(v_{nm}, v_{nk})) \leq \beta_{n+m,k-m-1}$$

we get for m=0 and $k \to \infty$ that

$$\rho_n(c_n, v_n) \leq \lim_{\nu} \beta_{n\nu} \leq \sup_{\nu} \beta_{n\nu} \to 0 \ (n \to \infty).$$

Next we show that v is a fixed point of T. The map T_n is continuous as is easily seen from (16.3). Therefore we are allowed to take the limit $k \to \infty$ in (16.6) which results in $v_n = T_n v_{n+1}$. Hence v is a fixed point of T. Finally we prove the uniqueness of v within B(c). For that purpose we consider an arbitrary fixed point $w \in T$ that belongs to B(c). Now we get

(16.7) $$\rho_n(w_n, v_n) = \rho_n(T_n w_{n+1}, T_n v_{n+1}) \leq \rho_{n+1}(w_{n+1}, v_{n+1}).$$

From (16.7) one derives by induction on k that

$$\rho_n(w_n, v_n) \leq \rho_{n+k}(w_{n+k}, v_{n+k})$$
$$\leq \rho_{n+k}(w_{n+k}, c_{n+k}) + \rho_{n+k}(c_{n+k}, v_{n+k}) \to 0 \quad (n \to \infty).$$

Therefore $w_n = v_n$, hence w=v. ⌋

Remarks. (i) From the proof one may extract the following estimate for the rate of convergence of $v_{nm} \to v_n$ $(m \to \infty)$:

$$\rho_n(v_n, v_{nm}) \leq \lim_{\nu} \ \beta_{n+m,\nu} \ .$$

(Note that we have started the iteration process for finding the fixed point in the special point c, while in the Banach fixed point theorem one may start in any point.)

(ii) One may wonder if theorem 16.1. really includes the fixed point theorem for contractions. In order to prove this we consider a complete metric space (D, ρ). Let $L: D \to D$ be a contraction with contraction modul $\alpha < 1$. Let d be an arbitrary point in D. For $m \in \mathbb{N}, n \in \mathbb{N}$ we define

$$B_n := D, \ \rho_n := \alpha^n \rho, \ T_n := L, \ \beta_{mn} := \rho(d, Ld) \cdot \alpha^m \cdot \sum_{o}^{n} \alpha^{\nu}, \ c_n := d.$$

The metric spaces (B_n, ρ_n) are complete and moreover $\sup_n \beta_{mn} = \rho(d, Ld)\alpha^m (1-\alpha)^{-1} \to 0$ $(m \to \infty)$. The map L obviously satisfies (16.3). By induction on k one can easily show that L also satisfies (16.4). Hence, according to theorem 16.1, there exists a sequence (v_n) of points $v_n \in D$ such that

(16.8) $$v_n = Lv_{n+1}\ ,\quad n \in \mathbb{N},$$

and (v_n) has the property

(16.9) $$\rho(v_m,d) \leq \alpha^{-m}\cdot \underline{\lim}_{n}\ \beta_{mn} = \rho(d,Ld)(1-\alpha)^{-1},\ m \in \mathbb{N}.$$

From (16.8) we conclude that $\rho(v_m,v_{m+1})=\rho(Lv_{m+1},Lv_{m+2}) \leq \alpha\rho(v_{m+1},v_{m+2})$, hence $\rho(v_m,v_{m+1})\leq\alpha^n\rho(v_{m+n},v_{m+n+1})$ for all $n \in \mathbb{N}$. The triangle inequality and (16.9) imply

$$\rho(v_m,v_{m+1}) \leq 2\alpha^n\rho(d,Ld)\cdot(1-\alpha)^{-1}\to 0 \quad (n\to\infty).$$

This shows that $v_1=v_n$ for all n, and (16.8) tells us that L has at least one fixed point. Since L is a contraction, it has also at most one fixed point. Thus we have shown that theorem 16.1 actually extends the fixed point theorem for contractions, which, of course, can be verified directly in a simpler manner. From (16.9) we also get the relation

$$\rho(v,w) \leq \rho(w,Lw)\cdot(1-\alpha)^{-1}$$

for the fixed point v and any point $w\in D$. This property of the fixed point may also be derived easily in a direct way.

We shall give applications of theorem 16.1 in section 17 and section 19.

17. Criteria of optimality and existence of $\bar{p}$-optimal plans.

The following two theorems constitute an analogue to theorem 3.8. Denote by $P_{n\pi}$ the distribution of χ_n under P_π. In particular, we have $P_{1\pi}=q_o=p$.

<u>Theorem 17.1</u>. If $-\infty<V<\infty$, then for any randomized plan π the following three statements are equivalent:

(i) π is $\bar{p}$-optimal,

(ii) $V_{1\pi} = V_1$ p-a.s.

(iii) $V_{n\pi} = V_n$ $P_{n\pi}$-a.s., $n \in \mathbb{N}$.

<u>Remark</u>. The implication (ii)⟹(iii) represents for n=2 the *principle of optimality*.

Proof. We know that $P_{n\pi} = q_o\pi_1 \dots q_{n-1}$, hence $P_\pi = P_{n\pi} \otimes (\pi_n q_n \dots)$, hence $V_\pi = E_\pi[\sum_1^{n-1} r_\nu \circ (\chi_\nu, \alpha_\nu)] + \int P_{n\pi}(dh) V_{n\pi}(h)$. It follows that $-\infty<V_\pi<\infty$ implies $-\infty<V_{n\pi}<\infty$ on a measurable set N_n^c of $P_{n\pi}$-measure one. The set $C_n := \{h \in \bar{H}_n : V_{n\pi}(h) < V_n(h)\}$ is u-measurable by theorem 13.2. Hence there exists a partition $C_n = B_n + N_n$, where B_n is measurable and where N_n is a set of $P_{n\pi}$-measure zero (we do not distinguish in the notation between $P_{n\pi}$ and its completion), hence $V_{n\pi}(h) < V_n(h)$ on B_n and $P_{n\pi}(B_n) = P_{n\pi}(C_n)$.

(i)⟹(ii): Assume π to be $\bar{p}$-optimal, but $p(C_1)>0$. Keeping in mind that $V \in \mathbb{R}$, we get $V = G_\pi = \int V_{1\pi} dp < \int V_1 dp$. The last term equals V by theorem 14.2, which is a contradiction.

(ii)⟹(iii): We proceed by induction on n. Let us assume that $P_{n\pi}(B_n)=0$. The set $Z := \{h \in \bar{H}_n : \pi_n q_n(h, (B_{n+1})_h) > 0\}$ is measurable. For $h \in ZN_n^c$ we get by means of the OE the relation

$$\begin{aligned} V_{n\pi}(h) &= \int \pi_n(h,da)[r_n(h,a) + \int q_n(h,a,ds) V_{n+1,\pi}(h,a,s)] \\ &< \int \pi_n(h,da)[r_n(h,a) + \int q_n(h,a,ds) V_{n+1}(h,a,s)] \\ &\leq U_n V_{n+1}(h) = V_n(h). \end{aligned}$$

Therefore $ZN_n^c \subset C_n$, hence $P_{n\pi}(Z)=0$. Now we get

$P_{n+1,\pi}(B_{n+1}) = \int_Z P_{n\pi}(dh) \pi_n q_n(h, (B_{n+1})_h) = 0$, hence $P_{n+1,\pi}(C_{n+1})=0$.

The implication (iii)⟹(ii) is trivial, the implication (ii)⟹(i) is easily proved. ┘

In order to get an analogous criterion for the $\overline{p}$-optimality of a deterministic plan, we need the following

Lemma 17.2. Let $g \in \Delta'$ be an H-plan, $f \in \Delta'$ be the plan defined by

$$f_n = g_n \circ h_{nf},\ n \in \mathbb{N}.$$

Then the distribution of χ_n under P_g is the same as the distribution of $h_{nf} \circ \eta_n$ under P_f.

Remark. Define a map $h_f: \Omega \to \overline{H}$ by means of

$$h_f((s_1,s_2,\ldots)) = (s_1,f_1(s_1),s_2,f_2(s_1,s_2),\ldots).$$

It is easy to see that h_f is $\mathcal{F}$-$\overline{\mathcal{H}}$-measurable. Then lemma 17.2 may be expressed in the following compact form: The distribution of h_f under P_f equals P_g.

Proof. Denote by $(P_g)\chi_n$ and $(P_f)h_{nf} \circ \eta_n$ the distribution whose equality is to be proved. This equality is true for n=1, because then both distributions equal q_o. Let us assume that the equality is true for some $n \in \mathbb{N}$. For $B \in \overline{\mathcal{H}}_n$, $C \in \mathcal{A}$ and $F \in \mathcal{T}$ we get

$$(17.1)\quad P_g(\chi_{n+1} \in B \times C \times F) = \int_B (P_g)_{\chi_n}(dh) q_n(h,g_n(h),F) \cdot 1_C(g_n(h)),$$

$$(17.2)\quad P_f(h_{nf} \circ \eta_n \in B \times C \times F\ ) =$$

$$= \int_{h_{nf}^{-1}(B)} (P_f)_{\eta_n}(dy) q_n(h_{nf}(y),f_n(y),F) 1_C(f_n(y)),$$

and the right-hand sides of (17.1) and (17.2) coincide by the induction hypothesis and a well-known theorem on transformations of integrals.$\rfloor$

Lemma 17.3. Let $(X,\mathcal{F})$, $(Y,\mathcal{G})$ be measurable spaces. Any measurable map $g: X \to Y$ is u-measurable.

Proof. We have to show that $g^{-1}(B)$ is u-measurable whenever B is u-measurable. Let μ be an arbitrary probability on $\mathcal{F}$ and ν its image under g. Then there exists a partition $B = B' + N$, where $B' \in \mathcal{G}$ and $N \subset N' \in \mathcal{G}$, $\nu(N') = 0$. It follows that $g^{-1}(B) = g^{-1}(B') + g^{-1}(N)$ and $g^{-1}(B') \in \mathcal{F}$, $g^{-1}(N) \subset g^{-1}(N') \in \mathcal{F}$, $\mu(g^{-1}(N')) = \nu(N') = 0$.
Therefore $g^{-1}(B)$ is μ-measurable for any μ, hence $g^{-1}(B)$ is u-measurable.$\rfloor$

Corollary 17.4. For any $f\in\Delta, n\in\mathbb{N}$, the map $G_n o h_{nf}: S^n \to \overline{\mathbb{R}}$ is u-measurable and $(G_n o h_{nf})(s_1,\ldots,s_{n-1},\cdot)$ is u-measurable in the last coordinate.

Theorem 17.5. Let P_{nf} denote the distribution of η_n under P_f. If $-\infty<G<\infty$ and if each of the reward functions r_n is bounded from above, then for any plan $f\in\Delta$ the following three statements are equivalent:

(i) f is $\overline{p}$-optimal,
(ii) $G_{1f} = G_1$ p-a.s.
(iii) $G_{nf} = G_n o h_{nf}$ P_{nf}-a.s., $n\in\mathbb{N}$.

Proof. According to lemma 15.3 there exists an H-plan $g\in\Delta'$ such that $f_n = g_n o h_{nf}$, $G_f = G_g$, and $G_{nf} = G_{ng} o h_{nf}$, $n\in\mathbb{N}$. From theorem 15.4 we conclude with the aid of lemma 17.2 the following equivalences:
f $\overline{p}$-optimal $\Leftrightarrow G_f = G = V = G_g \Leftrightarrow$ g $\overline{p}$-optimal;
$G_{1f} = G_1$ p-a.s. $\Leftrightarrow G_{1g} = G_1$ p-a.s.;
$G_{nf} = G_n o h_{nf}$ P_{nf}-a.s. $\Leftrightarrow G_{ng} o h_{nf} = G_n o h_{nf}$ P_{nf}-a.s. $\Leftrightarrow G_{ng} = G_n$ P_{ng}-a.s.. Now theorem 17.5 follows from theorem 17.1. ⌋

We are going to investigate how some of the results of section 5 carry over to standard DM's. As earlier, the operator L_n is defined by

$$L_n v(h,a) := r_n(h,a) + \int q_n(h,a,ds)\,v(h,a,s).$$

Here v may be in case (EN) an arbitrary map that is u-measurable in the last coordinate and bounded from above. We shall say that a property E, defined for the elements of a probability space $(X,\mathcal{F},P)$ holds P-a.s., if there is a set $B\in\mathcal{F}$ of P-measure one whose elements have the property E. It follows that the set B' of *all* $x\in X$ which have property E is measurable with respect to the completion P' of P and $P'(B')=1$.

Theorem 17.6 (*Criteria of optimality*, cp. theorems 5.1 and 5.2). Assume case (EN) and $G>-\infty$. For any plan $f\in\Delta$ the following statements are equivalent.
(α) f is $\overline{p}$-optimal;
(β) for any $n\in\mathbb{N}$ and for P_{nf}-almost all $y\in S^n$ the point

$f_n(y)$ is a maximum point of the map $a \to L_n G_{n+1}(h_{nf}(y),a)$;

(γ) the equality $G_n(h_{nf}(y)) = L_n G_{n+1}(h_{nf}(y), f_n(y))$ holds for any $n \in \mathbb{N}$ and for P_{nf}-almost all $y \in S^n$.

Proof. a) Let f be $\bar{p}$-optimal. For any $n \in \mathbb{N}$ the set $C_n := \{y \in S^n : G_{nf}(y) = G_n(h_{nf}(y))\}$ is u-measurable and has P_{nf}-measure one, by theorem 17.5. As in the proof of theorem 17.1 we may find a measurable set $F_n \subset S^n$ of P_{nf}-measure one on which $G_{nf}(y) = G_n(h_{nf}(y))$. From $1 = P_{n+1,f}(F_{n+1}) = \int P_{nf}(dy) q_{nf}(y, (F_{n+1})_y)$ we conclude, that the measurable set $\Phi_n := \{y \in S^n : q_{nf}(y, (F_{n+1})_y) = 1\}$ has also P_{nf}-measure one. Take some $y \in F_n \cap \Phi_n$ and put $b := h_{nf}(y)$. By means of the OE we get
$L_n G_{n+1}(b, f_n(y)) = \Lambda_{nf} G_{n+1,f}(y) = G_{nf}(y) = G_n(b) = U_n G_{n+1}(b)$,
hence $f_n(y)$ is a maximum point for all y in the measurable set $F_n \cap \Phi_n$ which has P_{nf}-measure one.

b) Let us asume that there is for any $n \in \mathbb{N}$ a measurable set $B_n \subset S^n$ of P_{nf}-measure one such that $f_n(y)$ is a maximum point of $L_n G_{n+1}(h_{nf}(y), \cdot)$ whenever $y \in B_n$. It follows by means of the OE that $\Lambda_{nf} G_{n+1} \circ h_{n+1,f} = G_n \circ h_{nf}$ holds on B_n. One can show by downward induction that there is a measurable set $B \subset S$ of p-measure one on which

$$G_1 = \Lambda_{1f} \Lambda_{2f} \cdots \Lambda_{nf} \, G_{n+1} \circ h_{n+1,f}.$$

Now it follows exactly as in the proof of theorem 5.1 that $G_1 \le G_{1f}$ on B, hence f is $\bar{p}$-optimal by theorem 17.5. Therefore (α) is equivalent to (β). The equivalence of (β) and (γ) follows from the OE (theorem 14.4) and from theorem 15.4. ⌋

There arises the question of characterizing $\bar{p}$-optimal randomized plans. In the search for an analogue of theorem 17.6(γ) we notice that its condition may be expressed in the form

$$\Lambda_{nf} G_{n+1} \circ h_{n+1,f} = G_n \circ h_{nf} \quad P_{nf}\text{-a.s.}, \ n \in \mathbb{N}.$$

In analogy to this condition we get

<u>Theorem 17.7</u>. Assume case (EN) and $G \succ -\infty$. The randomized plan π is $\bar{p}$-optimal iff for any $n \in \mathbb{N}$ the equation

$$\Lambda_{n\pi} G_{n+1} = G_n$$

holds $P_{n\pi}$-a.s.

The proof resembles that of theorem 17.6 and is therefore omitted.

In chapter I we could easily derive from theorem 5.1 (the analogue of theorem 17.6) the theorem 5.7 that tells us that there exists in case (EN) a $\bar{p}$-optimal plan if for any $n \in \mathbb{N}, h \in H_n$ the function $L_n G_{n+1}(h,\cdot)$ is upper semi-continuous and the set $D_n(h)$ is compact. Theorem 5.7 was then used in theorem 5.11, the main theorem of section 5. In order to get an analogue of theorem 5.7 we need an extension of a selection theorem of Dubins and Savage. We begin with some preparations.

The definition and some properties of upper semi-continuous (u.s.c.) maps $u:X\to \overline{\mathbb{R}}$ are given in section 5. In the remainder of this section the action space A is assumed to be compact. In this case the system of non-empty closed subsets of A is denoted by 2^A.

The set 2^A is a compact metric space under the so-called Hausdorff-metric (cf.Kuratowski (50)). We shall need the following

<u>Definition</u>. Let X be a standard Borel space. A map $D:X\to 2^A$ is called u.s.c. in the sense of Kuratowski (abbreviated by u.s.c.(K)) if $x_n, x\in X$, $a_n\in D(x_n)$, $x_n\to x$, $a_n\to a$ $(n\to\infty)$ implies $a\in D(x)$.

<u>Remark</u>. One easily proves the following fact:
Let D(x) be a non-empty subset of $A, x\in X$. Then $\{(x,a)\in X\times A: a\in D(x)\}$ is closed iff "D(x) is closed for any x and $D:X\to 2^A$ is u.s.c."

<u>Lemma 17.8</u>. We make the following assumptions:

(i) A is compact;

(ii) X is a standard Borel space;

(iii) D(x) is a non-empty subset of $A, x\in X$, such that $K:=\{(x,a)\in X\times A: a\in D(x)\}$ is closed.

(iv) $w:K\to \overline{\mathbb{R}}$ is u.s.c.

Put $W(x):=\{a\in D(x): w(x,a)= \sup_{a'\in D(x)} w(x,a')\}$.

Then $W:X\to 2^A$ is measurable.

Proof. a) Put $t(x):= \sup_{a\in D(x)} w(x,a)$. The proof of lemma 5.10 is valid under the assumptions of lemma 17.8 and tells us that $t:X\to \overline{\mathbb{R}}$ is u.s.c. Put $T:=\{(x,y)\in X\times \overline{\mathbb{R}}: t(x)\geq y\}$. Obviously

T is not empty. In particular, $(x,t(x))\in T$ $\forall x\in X$. T is also closed. In fact, consider a sequence of points $(x_n,y_n)\in T$ converging to (x,y). Then there are points $a_n\in D(x_n)$ such that $w(x_n,a_n)\geq y_n$. Since A is compact, there exists a subsequence (a_{n_k}) converging to some point $a\in A$. Then $(x_{n_k},a_{n_k})\to(x,a)\in K$, since K is closed. It follows that $y=\lim_n y_n\leq\overline{\lim_k}\, w(x_{n_k},a_{n_k})\leq w(x,a)$, hence $(x,y)\in T$.

b) Define the map $\tilde{W}:T\to 2^A$ by means of

$$\tilde{W}(x,y):=\{a\in D(x):w(x,a)\geq y\}.$$

Obviously $\tilde{W}(x,y)$ is always non-empty and closed, since w is u.s.c. Now we shall show that $\tilde{W}$ is u.s.c.(K). Let (x_n,y_n), $(x,y)\in T$; $a_n\in\tilde{W}(x_n,y_n)$, $(x_n,y_n)\to(x,y)$ and $a_n\to a$. We have to show that a belongs to $\tilde{W}(x,y)$. We have $y=\lim y_n\leq\overline{\lim_n}\, w(x_n,a_n)\leq w(x,a)$, hence $a\in\tilde{W}(x,y)$. It follows (cf. Kuratowski (50),p.38) that $\tilde{W}$ is measurable (if in T the trace-σ-algebra is used). The set T is closed, hence measurable. Therefore there exists a measurable extension of $\tilde{W}$ (which we also denote by $\tilde{W}$) from T to $X\times\overline{\mathbb{R}}$.

c) The map $t:X\to\overline{\mathbb{R}}$ is u.s.c., hence measurable. It follows that the map $x\to(x,t(x))$ from X into $X\times\overline{\mathbb{R}}$ is measurable, hence also $x\to\tilde{W}(x,t(x))=W(x)$ is measurable.⌋

Theorem 17.9. (Generalized selection theorem, cf.Dubins and Savage (65), chapter 2.16 and Maitra (68)).
Under the assumptions of lemma 17.8 there exists a measurable map $f:X\to A$ such that

$$(\alpha)\qquad f(x)\in D(x),\ x\in X,$$

$$(\beta)\qquad w(x,f(x))=\sup_{a\in D(x)} w(x,a),\ x\in X.$$

The proof is essentially that of Maitra (68).

a) It is well-known (cf.e.g.Dieudonné (60),p.134) that there exists a sequence (v_n) of continuous maps $v_n:A\to\mathbb{R}$ which separate points in A; i.e. for any pair $(a,b)\in A^2$ there exists some n such that $v_n(a)\neq v_n(b)$. For any $n\in\mathbb{N}$ let us define a map $V_n:2^A\to 2^A$ by $V_n(K):=\{a\in K:v_n(a)=\sup_{a'\in K} v_n(a')\}$. Maitra (68, lemma 3.3) has shown that V_n is measurable. Let W be as in lemma 17.8. Define $F_o:=W$, $F_n:=V_n\circ F_{n-1}$, $n\in\mathbb{N}$. It follows from lemma 17.8 that $F_n:X\to 2^A$ is measurable.

For any x the sequence of sets $F_n(x)$ is decreasing, since $V_n(K) \subset K$. Since A is compact, the set $F(x) := \bigcap_n F_n(x)$ is not empty. Fix $x \in X$ and consider two points $a,b \in F(x)$, hence $a,b \in F_n(x)$ $\forall n \in \mathbb{N}$. Therefore $v_n(a) = \max_{a' \in F_{n-1}} v_n(a') = v_n(b)$ $\forall n$. Since the sequence (v_n) is separating points, we have a=b, i.e. F(x) is a singleton, say $F(x) = \{f(x)\}$. Since $(F_n(x))$ is decreasing, we have $f(x) \in F_o(x) = W(x)$, consequently $f(x) \in D(x)$, and $w(x,f(x)) = \max_{a' \in D(x)} w(x,a')$, $x \in X$.

b) It is well known that for any $x \in X$ the sequence of sets $F_n(x)$, being a decreasing sequence of closed sets converges in the metric of 2^A to F(x) (cf.Kuratowski (48),p.245 and Kuratowski (50),p.21). The limit of a convergent sequence of measurable maps into a metric space is a measurable map (cf.Dynkin (61),p.17,18). Hence $F: X \to 2^A$ is measurable. For any closed subset B of A we have $\{x \in X : f(x) \in B\} = \{x \in X : F(x) \in \{B\}\}$, and the latter is a Borel set in X, since F is measurable and since the singleton {B} is a Borel set in the metric space 2^A. The σ-algebra of Borel sets in A is generated by the system of closed subsets. It follows that $f: X \to A$ is measurable. ┘

<u>Theorem 17.10</u>. (cp.theorem 5.7) Assume case (EN) and $G > -\infty$. There exists a $\bar{p}$-optimal (deterministic) plan if the following conditions are satisfied.

(i) A is compact;

(ii) the sets K_n are closed subsets of $H_n \times A$, $n \in \mathbb{N}$;

(iii) the maps $L_n G_{n+1} : K_n \to \bar{\mathbb{R}}$, $n \in \mathbb{N}$, are upper semi-continuous.

Proof. We apply theorem 17.9 by putting $X := H_n$, $D(x) := D_n(h)$, hence $K = K_n$, $w := L_n G_{n+1}$. According to theorem 17.9 there exists a measurable map $g_n : H_n \to A$ such that $g_n(h) \in D_n(h)$ and such that $g_n(h)$ is a maximum point of $L_n G_{n+1}(h,\cdot)$. Now $g = (g_n)$ is an H-plan, i.e. the sequence of maps $f_n := g_n \circ h_{nf}$ defines a plan $f \in \Delta$ such that $f_n(y)$ is a maximum point of $L_n G_{n+1}(h_{nf}(y),\cdot)$. Not it follows from theorem 17.6 that f is $\bar{p}$-optimal.

As in section 5 we have now the task to find sufficient conditions for the upper semi-continuity of $L_n G_{n+1}$ which are easily to check.

<u>Definition</u>. Let $(X,\mathcal{F})$ and $(Y,\mathcal{G})$ be SB-spaces. A transition probability q from X into Y is called *continuous*, if $x_n \to x$

implies that $q(x_n,\cdot)$ converges weakly to $q(x,\cdot)$, i.e.

$$\int q(x_n,dy)\varphi(y) \to \int q(x,dy)\varphi(y)$$

for any continuous and bounded map $\varphi: Y \to \mathbb{R}$.

Remark. In applications, Y will often be an euclidean d-space. In this case q is continuous iff $x_n \to x$ implies the convergence of the sequence of distribution functions $F_n(y) := q(x_n, (-\infty, y\rangle)$ to the distribution function $F(y) := \int q(x, (-\infty, y\rangle)$ at every point of continuity of the latter. In many cases q will be given by a density, i.e. in the form $q(x,B) = \int_B v(x,y)\lambda^d(dy)$, where $v \geq 0$ is measurable and where λ^d is the Lebesgue-measure in $\mathbb{R}^d$. If v is continuous in the first coordinate, then it follows from a theorem of Scheffé that q is continuous.

Lemma 4.1 in Maitra (68) carries easily over to

Lemma 17.11. Let $w: H_{n+1} \to \overline{\mathbb{R}}$ be u.s.c. and bounded, and let q_n be continuous. Then $\int q_n(\cdot, ds) w(\cdot, ds)$ is u.s.c. on K_n.

Now we are ready for the proof of the main result of this section.

Theorem 17.12. (cf. theorem 5.11). Let us assume that the following conditions are satisfied.

(i) The action space A is compact;
(ii) the sets K_n are closed subsets of $H_n \times A$;
(iii) the transition laws q_n are continuous;
(iv) case (C) holds;
(v) the reward functions are upper semi-continuous.

Then the maximal conditional expected reward G_n is upper semi-continuous and there exists a $\overline{p}$-optimal (deterministic) plan.

Proof. Denote the set of bounded u.s.c. functions on H_n by B_n. It follows from lemma 4.2 of Maitra (68) that B_n is complete under the metric $\rho_n(u,v) := \|u-v\|$. According to lemma 17.11, lemma 5.5 and lemma 5.10 the operator U_n, defined by

$$U_n w := \sup_{a \in D_n(h)} [r_n(\cdot,a) + \int q_n(\cdot,a,ds) w(\cdot,a,s)]$$

maps B_{n+1} into B_n. Now we shall apply the fixed point theorem 16.1. Put $\beta_{mn} := \sum_{i=m}^{m+n} \|r_i\|$, $m,n \in \mathbb{N}$. Then condition (16.2) is satisfied since $\sum\|r_i\|<a$. Let c_n be the identically vanishing function on H_n. Now we are going to verify (16.3) and (16.4). Using lemma 3.3 we get for $b,d\in B_{n+1}$

$$\rho_n(U_n b,U_n d)=\|\sup_{a\in D_n(\cdot)} L_n b(\cdot,a)-\sup_{a\in D_n(\cdot)} L_n d(\cdot,a)\| \leq$$

$$\leq \|\sup_{a\in D_n(\cdot)} |\int q_n(\cdot,a,ds)[b(\cdot,a,s)-d(\cdot,a,s)]|\| \leq \|b-d\|$$

$= \rho_{n+1}(b,d)$. This proves condition (16.3). In order to prove (16.4) we remark at first that

$$\|U_n b\| \leq \|r_n\| + \|b\|,$$

hence $\rho_m(c_m,U_m\ldots U_{m+n}c_{m+n+1})=\|U_m\ldots U_{m+n}0\|\leq \sum_m^{m+n}\|r_i\|=\beta_{mn}$. Therefore also condition (16.4) is satisfied. Now theorem 16.1 tells us that there exists one and only one sequence (v_n) of maps $v_n\in B_n$, such that $v_n=U_n v_{n+1}$, $n\in\mathbb{N}$, and $\lim_n\|v_n\|=0$. The maps v_n belong to the set $N_n^+\cap N_n^-$, hence they are bounded and universally measurable in the last coordinate. From theorem 14.6 and 15.4 we infer that $v_n=G_n$, hence G_n is u.s.c. Moreover, from lemma 17.11 and lemma 5.5 we conclude that the maps $L_n G_{n+1}$ are u.s.c. The existence of a $\overline{p}$-optimal plan is now a consequence of theorem 17.10. ⌋

18. Sufficient statistics, Markovian and stationary models.

Definition. Let $(F_n, \mathcal{F}_n)$ be SB-spaces, $n \in \mathbb{N}$. A sequence (t_n) of measurable maps $t_n : H_n \to F_n$ is called a *sufficient statistic* of the DM$(S,A,D,(p_n),(r_n))$, if the following conditions are satisfied:

(i) There exist non-empty sets $D_n'(t) \subset A$, $t \in F_n$, such that

$$K_n' := \{(t,a) \in F_n \times A : a \in D_n'(t)\} \tag{18.1}$$

is measurable and contains the graph of a measurable map;

$$D_n(h) = D_n'(t_n(h)), \quad n \in \mathbb{N}, \ h \in H_n. \tag{18.2}$$

(ii) There exist transition probabilities q_n' from K_n' into S,+) such that

$$q_n(h,a,\cdot) = q_n'(t_n(h),a,\cdot), \quad n \in \mathbb{N}, (h,a) \in K_n. \tag{18.3}$$

(iii) There exist measurable maps $r_n' : K_n' \to \overline{\mathbb{R}}$ such that

$$r_n(h,a) = r_n'(t_n(h),a), \quad n \in \mathbb{N}, \ (h,a) \in K_n. \text{++)} \tag{18.4}$$

(iv) There exist measurable maps $T_n : K_n' \times S \to F_{n+1}$ such that

$$t_{n+1}(h,a,s) = T_n(t_n(h),a,s), \quad n \in \mathbb{N}, \ (h,a,s) \in H_{n+1}. \tag{18.5}$$

Remarks. (i) In section 6 the maps t_n could without loss of generality be assumed to be surjective. Here this assumption may result in complications, since the range $t_n(H_n)$ of t_n need not belong to $\mathcal{F}_n$. Consequently the formulation of theorem 18.1 is a bit more complicated than that of theorem 6.0. Moreover, the maps D_n', q_n', r_n', T_n need not be uniquely determined by (D_n), (q_n), (r_n), and (t_n). They are uniquely determined "on $t_n(H_n)$".

(ii) For any DM there exists the trivial sufficient statistic $t_n(h) := h$, $n \in \mathbb{N}$, $h \in H$. Of course, one is interested in sufficient statistics where t_n "contracts" H_n substantially.

(iii) In pratical problems often maps D_n', q_n', r_n', t_n and T_n, which satisfy (18.1) and (18.5), are the original data, from which a DM is constructed by means of (18.2)-(18.4).

+) Whenever it is necessary, we extend q_n' (in an arbitrary way) to a transition probability from $F_n \times A$ into S, and r_n' will be extended by putting $r_n'(t,a) := 0$ for $(t,a) \in F_n \times A - K_n'$.

++) Moreover, we assume $\Sigma \| r_n'^{+} \| < \infty$ or $\Sigma \| r_n'^{-} \| < \infty$.

<u>Definition</u>. Let $t=(t_n)$ be a sufficient statistic of the DM.

a) A randomized plan $\pi\in\Delta^r$ is a *t-plan*, if there exist transition probabilities π'_n from F_n into A such that

$$\pi_n(h,\cdot) = \pi'_n(t_n(h),\cdot),\ n\in\mathbb{N}, h\in H_n.$$

b) A plan $f\in\Delta$ is a *t-plan*, if there exist measurable maps $\varphi_n:F_n\to A$ such that

$$f_n = \varphi_n \circ t_n \circ h_{nf},\ n\in\mathbb{N}.$$

<u>Theorem 18.1</u> (cp.Strauch (66) for Markovian models). Let $t=(t_n)$ be a sufficient statistic for the DM. Then there exists for any randomized plan π a randomized t-plan σ for which $G_\sigma=G_\pi$.

Proof. a) Fix some $n\in\mathbb{N}$. According to theorem 12.4 there exists a conditional P_π-distribution τ_n of α_n under the condition $t_n\circ\chi_n$. We shall prove that one can choose τ_n in such a way that

$$(18.6)\qquad \tau_n(t,D'_n(t)) = 1,\ t\in F_n.$$

We use corollary 12.7 by putting $X:=F_n$; $Y:=A$; $B:=K'_n$; q the transition probability determined by the measurable map whose existence is assumed in (18.1); $\mu:=(P_\pi)_{(t_n\circ\chi_n,\alpha_n)}$. The probability μ is concentrated on B, since $P_\pi((t_n\circ\chi_n,\alpha_n)\in K'_n)=\int(P_\pi)_{\chi_n}(dh)\pi_n(h,D_n(h))=1$. Now it follows from corollary 12.7 that we can assume that τ_n has the property (18.6). Therefore

$$(18.7)\qquad \sigma_n(h,\cdot) := \tau_n(t_n(h),\cdot),\ n\in\mathbb{N},\ h\in H_n,$$

defines a randomized plan $\sigma=(\sigma_n)$.

b) We show that $\tau_n\otimes q'_n$ is a conditional P_π-distribution of (α_n,ζ_{n+1}) under the condition $t_n\circ\chi_n$. In fact, if $B\in\mathfrak{F}_n$, $C\in\mathfrak{A}$, $L\in\mathfrak{T}$, then $\int_B(P_\pi)_{t_n\circ\chi_n}(dt)\int_C\tau_n(t,da)q'_n(t,a,L)=$

$$=\int_{B\times C}(P_\pi)_{(t_n\circ\chi_n,\alpha_n)}(d(t,a))q'_n(t,a,L)=$$

$$=\int_{t_n^{-1}(B)\times C}(P_\pi)_{(\chi_n,\alpha_n)}(d(h,a))q_n(h,a,L)=$$

$$=P_\pi(t_n\circ\chi_n\in B,\alpha_n\in C,\zeta_{n+1}\in L).$$

c) We assert that

(18.8) $$E_\pi W o(t_n o\chi_n,\alpha_n,\zeta_{n+1}) = E_\sigma W o(t_n o\chi_n,\alpha_n,\zeta_{n+1}).$$

holds for all $n\in\mathbb{N}$ and all measurable maps $W:F_n\times A\times S\to\overline{\mathbb{R}}$ which are bounded either from above or from below.
At first we look at the case n=1. We have $(P_\pi)_{t_1 o\zeta_1}=$ $=((P_\pi)_{\zeta_1})_{t_1}=(p)_{t_1}=(P_\sigma)_{t_1 o\zeta_1}$. Therefore
$E_\pi W o(t_1 o\zeta_1,\alpha_1,\zeta_2)=\int(P_\pi)_{t_1 o\zeta_1}(dt)\int\tau_1(t,da)\int q_1'(t,a,ds')W(t,a,s')=$
$=\int(P_\sigma)_{\zeta_1}(ds)\int\sigma_1(s,da)\int q_1(s,a,ds')W(t(s),a,s')=$
$=E_\sigma W o(t_1 o\zeta_1,\alpha_1,\zeta_2)$. Now let us assume that (18.8) holds for some $n\in\mathbb{N}$. Then $E_\pi W o(t_{n+1}o\chi_{n+1},\alpha_{n+1},\zeta_{n+2})=$
$=\int(P_\pi)_{(t_n o\chi_n,\alpha_n,\zeta_{n+1})}(dx)\int\tau_{n+1}(x,da)\int q_{n+1}(x,a,ds)W(T_n(x),a,s)=$
$=E_\pi W' o(t_n o\chi_n,\alpha_n,\zeta_{n+1})$, where
$$W'(x) := \int\tau_{n+1}(x,da)\int q_{n+1}(x,a,ds)W(x,a,s).$$

It follows from the induction hypothesis that
$E_\pi W o(t_{n+1}o\chi_{n+1}, \alpha_{n+1},\zeta_{n+2})=E_\sigma W' o(t_n o\chi_n,\alpha_n,\zeta_{n+1})=$
$=E_\sigma W o(t_{n+1}o\chi_{n+1},\alpha_{n+1},\zeta_{n+2})$. Theorefore (18.8) is true for all $n\in\mathbb{N}$.
d) From theorem A4 we conclude that
$G_\pi=E_\pi(\sum r_n o(\chi_n,\alpha_n))=\sum E_\pi r_n o(\chi_n,\alpha_n)=\sum E_\pi r_n' o(t_n o\chi_n,\alpha_n)=$
$=\sum E_\sigma r_n' o(t_n o\chi_n,\alpha_n)=G_\sigma$. ┘

<u>Remarks</u>. In contrast to theorem 15.2 there does not always exist for any randomized plan σ a randomized t-plan σ, such that $V_{1\sigma}\geq V_{1\pi}$. This may be shown by means of an example given by Blackwell (65) (cf. the example after theorem 14.5; cp. also Strauch (66)).

<u>Theorem 18.2</u>. Let $t=(t_n)$ be a sufficient statistic, and assume case (EN). Then there exists for any randomized plan a deterministic t-plan f such that $G_f \geq G_\pi$.

Proof. At first we conclude from theorem 18.1 that there exists a randomized t-plan σ for which $G_\sigma=G_\pi$. Now we are going to use the ideas in the proof of theorems 9.4 and 15.2. Let R_n^- be the set of measurable maps $v:F_n\to\overline{\mathbb{R}}$ which are bounded from above. For the t-plan σ we define the operator

$$\Lambda_n' v := \int\sigma_n'(\cdot,da)\left[r_n'(\cdot,a)+\int q_n'(\cdot,a,ds)v(T_n(\cdot,a,s))\right],$$

which maps R_{n+1}^- into R_n^-. It is easily seen that

(18.9) $$(\Lambda_{n\sigma}' v)ot_n = \Lambda_{n\sigma}(vot_{n+1}).$$

It follows from (18.9) by induction on k that

(18.10) $\quad \Lambda_{n\sigma}\ldots\Lambda_{n+k,\sigma}0 = (\Lambda'_{n\sigma}\ldots\Lambda'_{n+k,\sigma}0)\circ t_n, \; k\in\mathbb{N}_o.$

Assumption (EN) implies the convergence of $\Lambda'_{n\sigma}\ldots\Lambda'_{n+k,\sigma}0$ $(k\to\infty)$ to some function $V'_{n\sigma}\in R^-_n$, hence

$$V_{n\sigma} = V'_{n\sigma}\circ t_n$$

by (18.10) and lemma 11.2. The map

$$v'_n := r'_n + \int q'_n(\cdot,\cdot,ds)V'_{n+1,\sigma}(\cdot,\cdot,s)$$

is defined on $F_n\times A$, and it is measurable and bounded from above. Moreover, one shows as in the proof of lemma 3.5. that

$$V'_{n\sigma} = \int\sigma'_n(\cdot,da)v'_n(\cdot,a).$$

As in theorem 15.2 we conclude by means of lemma 15.1 that there exist measurable maps $\varphi_n:F\to A$, such that for $n\in\mathbb{N}, t\in F_n$

(18.11) $\quad \varphi_n(t)\in D'_n(t)$

(18.12) $\quad v'_n(t,\varphi_n(t)) \geq V'_{n\sigma}(t)$.

From (18.11) we see that the sequence $g:=(g_n)$ defined by $g_n:=\varphi_n\circ t_n$, is an H-plan (cf.section 11 for the definition). From (18.12) we get $v_n(h,g_n(h))=v'_n(t_n(h),\varphi_n(t_n(h))) \geq$ $\geq V'_{n\sigma}(t_n(h))=V_{n\sigma}(h)$. Now it follows exactly as in the proof of theorem 9.4 that $G_g\geq G_\sigma$. Finally we put $f_n:=g_n\circ h_{nf}$, $n\in\mathbb{N}$. Then f is a t-plan, and $G_f=G_g\geq G_\sigma$ (cf.(9.11)).$\_|$

Corollary 18.3. Let t be a sufficient statistic. If there exists a randomized $\bar{p}$-optimal plan, then there exists a randomized $\bar{p}$-optimal t-plan σ. In case (EN) the randomized plan σ may be replaced by a deterministic plan.

Denote $t_n(H)$ by F'_n. Let Q^+_n and Q^-_n be the set of maps $v:F'_n\to\overline{\mathbb{R}}$, such that $v\circ t_n\in N^+_n$ or $v\circ t_n\in N^-_n$, respectively (for the definition of N^+_n and N^-_n cf.section 14). The operator U'_n, defined by

(18.13) $\quad U'_n v := \sup_{a\in D'_n(\cdot)} [r'_n(\cdot,a)+\int q'_n(\cdot,a,ds)v(T_n(\cdot,a,s))],$

is defined on Q^+_{n+1} and Q^-_{n+1} in case (EP) and (EN),respectively. $U'_n v$ need not belong to Q^+_n or Q^-_n.

Definition. Let (t_n) be a sufficient statistic of the DM. A solution of the *reduced optimal equation* is a sequence (v_n) of maps $v_n \in Q_n^+$ (in case (EP)) or $v_n \in Q_n^-$ (in case (EN)) such that

$$v_n = U_n' v_{n+1}, \quad n \in \mathbb{N}.$$

Theorem 18.4. (cf. theorem 6.0) Let $t=(t_n)$ be a sufficient statistic, let Δ^t be the set of t-plans. Then the following statements hold.

(i) $V_n = \sup_{\pi \in \Delta^t} V_{n\pi}$, $n \in \mathbb{N}$.

(ii) There exists for any $n \in \mathbb{N}$ a unique map $V_n' : F_n' \to \overline{\mathbb{R}}$ such that

$$V_n = V_n' \circ t_n.$$

(V_n') is a solution of the reduced optimality equation.

(iii) In case (EP) the sequence (V_n') is the termwise smallest of those solutions (v_n) of the reduced OE that satisfy

$$\text{(18.14)} \qquad \lim_n \|v_n^-\| = 0.$$

(iv) In case (C) the sequence (V_n') is the unique solution (v_n) of the reduced OE that satisfies

$$\text{(18.15)} \qquad \lim_n \|v_n\| = 0.$$

Proof. (i) Let π be an arbitrary randomized plan. Fix some $x \in S$ and take for p the measure concentrated at $\{x\}$. Assume $G<\infty$. According to theorem 18.1 and theorem 14.2 there exists for any $\varepsilon>0$ a t-plan π such that $V_{1\pi}(x)=G_\pi \geq G-\varepsilon = V_1(x)-\varepsilon$. If $G=\infty$, then there exists for any $m \in \mathbb{N}$ a t-plan π such that $V_{1\pi}=G_\pi \geq m$. It follows that (i) is true for n=1. In order to prove (i) for m>1, we use the standard reduction method of section 13. Fix some $k=(h,a) \in K_m$ and denote the elements in the shifted $\overline{DM}$ by a bar, e.g. $\overline{D}_n(h) := D_{m+n}(k,h)$. We are going to prove that the sequence of maps $\overline{t}_n : (H_{m+n})_k \to F_{m+n}$, defined by

$$\text{(18.16)} \qquad \overline{t}_n(h) := t_{m+n}(k,h), \quad n \in \mathbb{N}, h \in (H_{m+n})_k,$$

is a sufficient statistic for the model $\overline{DM}$. In fact, if we put $\overline{F}_n := F_{m+n}$, $\overline{D}_n'(t) := D_{m+n}'(t)$, $\overline{q}_n' := q_{m+n}'$, $\overline{r}_n' := r_{m+n}'$, $\overline{T}_n := T_{m+n}$, $n \in \mathbb{N}$, then it is easily seen that the conditions (18.1)-(18.5) are satisfied. It follows that $V_{m+1} = \overline{V}_1 = \sup_{\pi \in \overline{\Delta}^t} \overline{V}_{1\pi} = \sup_{\pi \in \Delta^t} V_{m+1,\pi}$, hence (i) is true for n=m+1.

(ii) Fix $n \in \mathbb{N}$, $h,h' \in H_n$ such that $t_n(h)=t_n(h')$. For any $\pi \in \Delta^t$ we get $(\Lambda_{n\pi} v)(h) = (\Lambda_{n\pi} v)(h')$, whenever $v \in M_{n+1}^+$ (in case (EP)) or $v\in M_{n+1}^-$ (in case (EN)) has the property that $v(h,\cdot)=v(h',\cdot)$. It follows by induction on k that $(\Lambda_{n\pi}\cdots\Lambda_{n+k,\pi}0)(h)=$ $=(\Lambda_{n\pi}\cdots\Lambda_{n+k,\pi}0)(h')$, hence $V_{n\pi}(h)=V_{n\pi}(h')$ by lemma 11.2. From part (i) we conclude that $V_n(h)=V_n(h')$, which proves the existence of a unique map $V_n':F_n'\to\overline{\mathbb{R}}$ such that $V_n=V_n'\circ t_n$. That (V_n') is a solution of the reduced OE is now easily derived from theorem 14.4. Part (iii) and part (iv) are derived from theorem 14.6 in a similar way. ┘

The following theorem and its proof are completely analogous to that of the corresponding part of theorem 6.0.

Theorem 18.5. Let (t_n) be a sufficient statistic and assume case (EP). Then there is well-defined a double-sequence (V_{nk}') of maps $V_{nk}':F_n'\to\overline{\mathbb{R}}$ by

$$V_{nk}' := U_n'V_{n+1,k-1}', \quad n\in\mathbb{N}, k\in\mathbb{N},$$

$$V_{no}' :\equiv 0 ,$$

and we have $V_n'=\lim_k V_{nk}'$, $n\in\mathbb{N}$.

Remark. It is not clear a priori that (V_{nk}') exists, since U_n' is only defined on Q_n^+. However, one easily shows by induction on k, that $V_{nk}'\circ t_n=V_{nk}$ which implies $V_{nk}'\in Q_n^+$, since V_{nk} belongs to N_n^+ by theorem 14.5.

Let us say a few words to Markovian models.

Definition. a) A DM is called *Markovian* if the sequence $\hat{t}$ of maps $t_n(h):=s_n, n\in\mathbb{N}, h\in H_n$, is a sufficient statistic. Hence $F_n=F_n'=S$ and $T_n(h,a,s)=s$.
b) A *randomized Markov plan* is a randomized $\hat{t}$-plan, a *Markov plan* is a $\hat{t}$ - plan.

It follows that a Markovian DM is determined by sets $D_n(s)$ +), transition laws $p_n(s,\cdot)$ and reward functions $r_n(s,a)$, such that $K_n':=\{(s,a)\in S\times A: a\in D_n(s)\}$ is measurable and contains the graph of a measurable map. We have $Q_n^+=M_1^+$, $Q_n^-=M_1^-$, and the operator U_n', defined by (18.13) takes on the form

$$U_n'v: = \sup_{a\in D_n(\cdot)} [r_n(\cdot,a)+\int q_n(\cdot,a,dj)v(j)].$$

+) We omit here the dash used earlier in the description of models which have an arbitrary sufficient statistic.

We do not intend to specialize theorems 18.1-18.5 to Markovian models, since this would result in a mere repetition. However, we shall show that the results, derived in section 6 for stationary models, remain true under our general assumption that the state space and the action space are SB-spaces.

Definition. The Markovian model $(S,A,D,(p_n),(r_n))$ is called stationary if D_n and p_n $(n\in\mathbb{N})$ do not depend on n, and if r_n is of the form $r_n=\beta^{n-1}r$ for some (measurable) function r and some $\beta\in(0,1\rangle$.

For stationary models we have the following equivalences.
Case (EN): either "$r\leq 0$" or "$\beta<1$ and r bounded from above";
Case (EP): either "$r\geq 0$" or "$\beta<1$ and r bounded from below";
Case (C) : (*discounted case*): $\beta<1$ and r bounded.

At first we get the following counterpart to theorem 6.7.

Theorem 18.6. (cp.Blackwell (65), Strauch (66)).
For any stationary model we have:
(i) The sequence (V_n) of maximal expected rewards satisfies

$$(18.17) \qquad V_n(h) = \beta^{n-1}V_1(s_n),\ n\in\mathbb{N}, h\in H_n.$$

(ii) V_1 is a solution v of the OE

$$(18.18) \qquad v = U_1(\beta v),\ \text{i.e.}$$

$$v= \sup_{a\in D(\cdot)} \left[r(\cdot,a)+\beta\int q(\cdot,a,ds)v(s)\right].$$

(iii) If $r\geq 0$, then V_1 is the smallest positive solution of (18.18). If r is bounded from below and $\beta<1$, then V_1 is the smallest of those solutions v of (18.18) for which $v\geq -\|r^-\|/(1-\beta)$. If r is bounded and $\beta<1$, then V_1 is the unique bounded solution of (18.18).
(iv) The sequence $(v_k,\ k\in\mathbb{N}_o)$ is well defined by

$$v_o: \equiv 0\ ,$$
$$v_k: = U_1(\beta v_{k-1}),\ k\in\mathbb{N}\ ,$$

and (v_k) converges to G_1 in case (EP).

Proof. (i) Let π be a randomized Markov plan. Denote by $^n\pi$ the randomized Markov plan $(\pi_n,\pi_{n+1},\ldots)$. Let $v:H_{n+k}\to\overline{\mathbb{R}}$ be measurable and bounded from above (in case (EN)) or from below (in case (EP)). One easily shows by induction on k that $(\Lambda_{n\pi}\cdots\Lambda_{n+k,\pi}v)(h)=\beta^{n-1}(\Lambda_{1\,{}^n\pi}\cdots\Lambda_{k\,{}^n\pi}v)(s_n)$. It follows from lemma 11.2 that $V_{n\pi}(h)=\beta^{n-1}V_{1\,{}^n\pi}(s_n)$. The map $\pi\to{}^n\pi$ from the set Δ^m of randomized Markov plans into Δ^m is obviously surjective. Hence part (i) of theorem 18.4 implies $V_n(h)=\sup_{\pi\in\Delta^m} V_{n\pi}(h)=\beta^{n-1}\cdot\sup_{\pi\in\Delta^m}V_{1\,{}^n\pi}(s_n)=\beta^{n-1}\sup_{\sigma\in\Delta^m}V_{1\sigma}(s_n)=$ $=\beta^{n-1}\cdot V_1(s_n)$. - The parts (ii)-(iv) are now easily derived from part (i) and from theorems 18.4 and 18.5. ⌋

<u>Remark</u>. We know from theorem 15.4 that $V_1=G_1$ whenever r is bounded from above.

<u>Theorem 18.7</u> (cp. theorem 6.9) Assume $G>-\infty$ and either "$r\leq 0$" or "r bounded from above and $\beta<1$". Then for any stationary plan f the following statements are equivalent.

(α) f is $\bar{p}$-optimal;

(β) f(s) is a maximum point of the map

$$a\to r(s,a)+\beta\int q(s,a,dj)G_1(j) \tag{18.19}$$

for all points s of a measurable set $Z\subset S$ for which

$$P_f(\zeta_n\in Z) = 1,\ n\in\mathbb{N}; \tag{18.20}$$

(γ) The equality

$$G_1(s) = r(s,f(s))+\beta\int q(s,f(s),dj)G_1(j)$$

holds for all points s of a measurable set $Z\subset S$ for which (18.20) holds.

Proof. a) Let f be $\bar{p}$-optimal. According to theorem 17.6 there exist measurable sets $Z_n\subset S^n$ such that $f(s_n)$ is a maximum point of the map (18.19) whenever $y=(s_1,\dots,s_n)\in Z_n$. The projection W_n of Z_n into the last factor of S^n is universally measurable by corollary 12.9. It follows (cf. the proof of lemma 12.1) that W_n contains a measurable set W_n' such that $P_f(\zeta_n\in W_n')=P_f(\zeta_n\in W_n)\geq P_f(\eta_n\in Z_n)=1$. The set $Z:=\bigcup_n W_n'$ has the property stated in (β).

b) Assume that (β) is true, and put $Z_n:=S^{n-1}\times Z$. Then $P_f(\eta_n\in Z_n)=P_f(\zeta_n\in Z)=1$, and $f(s_n)$ is a maximum point of the map (18.19) whenever $y\in Z_n$. Hence f is $\bar{p}$-optimal by theorem 17.5. c) The equivalence of (β) and (γ) is derived from the OE (theorem 18.6), since $G_1=V_1$ by theorem 15.4. ⌋

Theorem 18.8 (cp.Blackwell (65), Strauch (66)).
Assume $G>-\infty$ and either "$r\leq 0$" or "r bounded from above and $\beta<1$". If there exists a $\bar{p}$-optimal randomized plan then there exists a stationary deterministic $\bar{p}$-optimal plan.

Proof. We know from corollary 18.3 that if there exists a $\bar{p}$-optimal randomized plan, then there exists a deterministic Markov plan $f=(f_n)$. According to theorem 17.5 there exists a measurable set $B_1\subset S$ with $p(B_1)=1$ and a measurable set $B_2\subset S^2$ with $P_f(\eta_2\in B_2)=1$ such that $G_{1f}=G_1$ on B_1 and $G_{2f}=G_2\circ h_{2f}$ on B_2. It follows by means of part (i) of theorem 18.6 and its proof that $G_{2f}(s_2)=\beta G_1(s_2)$ whenever $(s_1,s_2)\in B_2$. It follows from $1=P_f(\eta_2)=$ $=\int p(ds_1)q(s_1,f_1(s_1),(B_2)_{s_1})$, that $q(s_1,f_1(s_1),(B_2)_s)=1$ for all s_1 in a measurable set $B_o\subset S$ with $p(B_o)=1$. Hence if $s\in B_o\cap B_1$, we get $G_1(s)=G_{1f}(s)=r(s,f(s))+\beta\int q(s,f(s),dj)G_1(j)$. Since $p(B_o\cap B_1)=1$, we conclude fom theorem 18.7, that the stationary plan $(f_1,f_1,\ldots)$ is $\bar{p}$-optimal.

Remark. Blackwell (65) and Strauch (66) prove the statement obtained from theorem 18.8 by replacing "$\bar{p}$-optimal" by "optimal". Note that the conclusion as well as the assumption of that statement is stronger than in theorem 18.8. hence both statements cannot be compared.

From theorems 17.12 and 18.8 we get immediately

Theorem 18.9 (cf. Maitra (68)). Let the model be stationary. Let us assume

α) the action space A is compact;

β) the set $K:=\{(s,a)\in S\times A: a\in D(s)\}$ is a closed subset of $S\times A$;

γ) the transition law q is continuous, i.e. if $(s_n,a_n)\to(s,a)$, then $q(s_n,a_n,\cdot)$ converges weakly to $q(s,a,\cdot)$;

δ) r is upper semi-continuous and bounded, $\beta<1$.

Then V_1 is upper-semi-continuous and there exists a $\bar{p}$-optimal stationary plan.

Remark. Maitra (68) has shown that under the conditions α),γ),δ) and β'): $K=S\times A$, there exists even an optimal stationary plan. It is possible to replace β') by β), using the ideas in section 17.

19. Validity of the optimality equation without topological assumptions on state space and action space.

The work of Blackwell (65) and Strauch (66) made it clear that a nice theory for DM's requires some topological assumptions on the nature of the σ-algebra in S and A. In practical problems these assumptions should always be fulfilled. However, some results in Blackwell (65) do in fact not require any topological assumption and may be generalized to our general model. This fact was shown in Hinderer (67) and will be reproduced here.

We shall use the DM introduced in section 11 *without* the assumption that S and A are SB-spaces.

Definition. (cp.Blackwell (65)). Let $\delta=(\delta^i)$ be a sequence of randomized plans $\delta^i\in\Delta^r$. We say that $\pi\in\Delta^r$ is δ-*generated*, if there exists for any $n\in\mathbb{N}$ a measurable partition

$$H_n = \sum_{i\in\mathbb{N}} B_{ni}\ ,\quad B_{ni}\in\mathcal{H}_n\ ,$$

such that π_n equals δ_n^i on $B_{ni}\times\mathcal{A}$, $i\in\mathbb{N}$.

We define (cp.Blackwell (65)) operators $T_{n\delta}$ by

$$(19.1)\qquad T_{n\delta}v := \sup_{i\in\mathbb{N}} \Lambda_{n\delta^i}v,\ n\in\mathbb{N}.$$

$T_{n\delta}$ is defined on M_{n+1}^- in case (EN) and on M_{n+1}^+ in case (EP).

Lemma 19.1. (cp.Blackwell (65)). Let $\delta=(\delta^i)$ be a sequence of randomized plans, and assume case (EN).

(i) For any δ-generated plan π we have

$$\Lambda_{n\pi}u \le T_{n\delta}u,\ n\in\mathbb{N},\ u\in M_{n+1}^- .$$

(ii) Let (u_n) be a sequence of functions $u_n\in M_n^-$ and (ε_n) a sequence of positive real numbers. Then there exists a δ-generated plan π such that

$$\Lambda_{n\pi}u_{n+1} \ge T_{n\delta}u_{n+1}-\varepsilon_n\ ,\quad n\in\mathbb{N}.$$

Proof. a) Let $H_n=\sum_i B_{ni}$ be the partition of H_n such that $\pi_n=\delta_n^i$ on $B_{ni}\times\mathcal{A}$. It follows immediately that $h\in B_{ni}$ implies $\Lambda_{n\pi}u(h)=\Lambda_{n\delta^i}u(h) \le T_{n\delta}u(h)$.

b) Put $B_{ni}:=\{h\in H_n:\ \Lambda_{n\delta^\nu}u_{n+1}<T_{n\delta}u_{n+1}-\varepsilon_n$ for $1\le\nu<i$,

$$\Lambda_{n\delta^i}u_{n+1}\ge T_{n\delta}u_{n+1}-\varepsilon_n\}.$$

The set B_{ni} belongs to $\mathcal{G}_n$ and furthermore $H_n = \sum_i B_{ni}$, since $T_{n\delta} u_{n+1} \leq \|r_n^+\| + \|u_{n+1}^+\| < \infty$. If we define $\pi_n(h,\cdot) : \delta_n^i(h,\cdot)$ for $h \in B_{ni}$, $n \in \mathbb{N}$, $i \in \mathbb{N}$, then (π_n) has the desired property. ┘

Theorem 19.2. (cf. Hinderer (67)). Assume case (C). Let $\delta = (\delta^i)$ be a sequence of randomized plans, and define

$$W_n := \sup_\pi G_{n\pi},$$

where the supremum is taken over the set of δ-generated plans. Then we have:

(i) For any $\varepsilon > 0$ there exists a δ-generated plan π such that

$$G_{n\pi} \geq W_n - \varepsilon, \; n \in \mathbb{N}.$$

(ii) The sequence (W_n) is the unique solution (u_n) of the equation

$$u_n = T_{n\delta} u_{n+1}, \; n \in \mathbb{N}$$

which satisfies

$$\lim_n \|u_n\| = 0. \tag{19.2}$$

(iii) The map $W_n : H_n \to \mathbb{R}$ is measurable.

Proof. At first we show by means of theorem 16.1 that $T := (T_{n\delta})$ has a unique fixed point (u_n) that satisfies (19.2). Let M_n be the set of measurable and bounded functions $u : H_n \to \mathbb{R}$. It is well-known that M_n is a Banach space under the norm $\|u\| := \sup_{h \in H_n} |u(h)|$. (This property follows from the fact, that M_n is a closed subspace of the Banach space of all bounded functions from H_n to $\mathbb{R}$.) Now we shall apply theorem 16.1 with $\rho_n(u,v) := \|u-v\|$, $u,v \in M_n$ and $\beta_{mn} := \sum_{k=m}^{m+n} \|r_k\|$. Condition (16.2) is true, since $\sum \|r_k\| < \infty$. Let c_n be the element of M_n that is identically zero. Now we are going to verify the conditions (16.3) and (16.4). Let us write T_n instead of $T_{n\delta}$. Using lemma 3.3 we get

$$\begin{aligned} \rho_n(T_n b, T_n d) &= \|T_n b - T_n d\| = \sup_h \left| \sup_i \Lambda_{n\delta^i} b - \sup_i \Lambda_{n\delta^i} d \right| \\ &\leq \sup_{(h,i)} \left| \int \delta_n^i(h,da) \int q_n(h,a,ds) [b(h,a,s) - d(h,a,s)] \right| \\ &\leq \|b-d\| = \rho_{n+1}(b,d). \end{aligned}$$

This proves condition (16.3). In order to prove (16.4) we remark at first that we get for any $u \in M_{n+1}$

(19.3) $\qquad \|T_n u\| \leq \|r_n\| + \|u\|.$

Applying (19.3) n times we get

$\rho_m(c_m, T_m \ldots T_{m+n} c_{m+n+1}) = \|T_m \ldots T_{m+n} 0\| \leq \sum_m^{m+n} \|r_k\| = \beta_{mn}.$

Hence (16.4) is true, and theorem 16.1 tells us that there exists exactly one sequence (u_n) of maps $u_n \in B_n$ such that

(19.4) $\qquad u_n = T_{n\delta} u_{n+1}, n \in \mathbb{N},$

and

(19.5) $\qquad \lim_n \|u_n\| = 0.$

Now we are going to show that

(19.6) $\qquad G_{n\pi} \leq u_n, \ n \in \mathbb{N}$

holds for any δ-generated plan π. Let us write Λ_n instead of $\Lambda_{n\pi}$. By a repeated application of the formula $\|\Lambda_n u - \Lambda_n v\| \leq \|u-v\|$ we get $\|\Lambda_n \ldots \Lambda_{n+m} u_{n+m+1} - \Lambda_n \ldots \Lambda_{n+m} 0\| \leq \|u_{n+m+1}\|$. From (19.5) and lemma 11.2 we conclude that $\Lambda_n \ldots \Lambda_{n+m} u_{n+m+1}$ converges to $G_{n\pi}$ for $m \to \infty$. Finally lemma 19.1 and (19.4) imply $\Lambda_n \ldots \Lambda_{n+m} u_{n+m+1} \leq T_n \ldots T_{n+m} u_{n+m+1} = u_n$, and (19.6) is verified. Therefore $W_n \leq u_n, n \in \mathbb{N}$, and the proof of theorem 19.2 will be complete as soon as we have proved that there exists for any $\varepsilon > 0$ a δ-generated plan π such that

(19.7) $\qquad G_{n\pi} \geq u_n - \varepsilon, \ n \in \mathbb{N}.$

Now there exists according to lemma 19.1 a δ-generated plan π such that $\qquad \Lambda_{n\pi} u_{n+1} \geq T_n u_{n+1} - \varepsilon . 2^{-n}, n \in \mathbb{N}.$

Write again Λ_n instead of $\Lambda_{n\pi}$. Λ_n is isotone and has the property that $\Lambda_n(v+\alpha) = \Lambda_n v + \alpha$ for any constant $\alpha \in \mathbb{R}$. Therefore we get $\Lambda_n \ldots \Lambda_{n+m} u_{n+m+1} \geq T_n \ldots T_{n+m} u_{n+m-1} - \sum_n^{n+m} \varepsilon . 2^{-\nu} \geq u_n - \varepsilon$ for any $m \in \mathbb{N}$. Some lines above we have shown that $\Lambda_n \ldots \Lambda_{n+m} u_{n+m+1}$ converges to $G_{n\pi}$, which completes the proof of (19.7). $\rfloor$

<u>Remark</u>. One may replace the number ε in part (i) of the theorem by any sequence (ε_n) of positive numbers (cf. Hinderer (67)).

<u>Definition</u>. A randomized plan π is called strongly ε-optimal if $G_{n\pi} \geq V_n - \varepsilon, n \in \mathbb{N}$.

<u>Theorem 19.3</u> (cf.Hinderer (67)). Let the state space $(S,\mathcal{T})$ and the action space $(A,\mathfrak{A})$ be arbitrary measurable spaces. Let us assume that there exists for any $\varepsilon>0$ a strongly ε-optimal randomized plan. Let us assume that case (C) holds. Then we have:

(i) The maximal conditional expected reward V_n is $\mathcal{G}_n$-measurable.

(ii) The sequence (V_n) is the unique solution (v_n) of the OE

$$v_n = U_n\, v_{n+1},\ n \in \mathbb{N}\ ,$$

for which

$$\lim_n \|v_n\| = 0. \tag{19.8}$$

<u>Remark</u>. Theorem 19.3 has been proved by Blackwell (65) for the stationary discounted model. Sirjaev (67) states part (ii) of the theorem (within the context of models with incomplete information, cf.section 7), but in his proof the measurability of V_n is tacitly assumed.

Proof. (i) Fix some $n \in \mathbb{N}$. Let δ^i be a strictly i^{-1}-optimal plan, $i \in \mathbb{N}$, and let $T_{n\delta}$ be the operator defined in (19.1) by means of the sequence $\delta=(\delta^i)$. Let (W_n) be as in theorem 19.2, hence $W_n \geq G_{n\delta^i} \geq V_n - i^{-1}, i \in \mathbb{N}$, therefore $W_n \geq V_n$. According to part (i) of theorem 19.2 there exists for any $m \in \mathbb{N}$ a δ-generated plan π^m such that $V_n \geq G_{n\pi^m} \geq W_n - m^{-1}$. Therefore also $V_n \geq W_n$, hence V_n equals the function W_n which is measurable according to part (iii) of theorem 19.2. - Part (ii) of the theorem is proved quite similarly as theorem 14.6. ⌋

20. Supplementary remarks.

In this section we supply some additional information on the foundation of dynamic programming.

A. Notions of optimality.

We generalize our decision model of section 11 by replacing the set Δ^r of randomized plans by some given non-empty subset $C \subset \Delta^r$, which we call the *set of admissible plans*. (Examples: C=set Δ' of H-plans, C=set of all Markov plan, C=set of all randomized Markov plans.) By choosing $D_n(h):=A$, we can describe by C the most general situation. Then we define

$$W: = \sup_{\pi \in C} V_\pi, \qquad W_1: = \sup_{\pi \in C} V_{1\pi}.$$

Definition. Let γ,ε be real numbers, $0 \leq \gamma < 1$, $\varepsilon \geq 0$. Let π be a randomized plan in C.

a) π is called *(γ,p,ε)-optimal*, if

$$(20.1) \qquad p(V_{1\pi} \geq V_{1\delta} - \varepsilon) \geq 1-\gamma \qquad \forall \delta \in C.$$

π is called (p,ε)-optimal or (γ,p)-optimal or *p-optimal*, if it is $(0,p,\varepsilon)$-optimal or $(\gamma,p,0)$-optimal or $(0,p,0)$-optimal, respectively.

b) π is called *ε-optimal*, if

$$(20.2) \qquad V_{1\pi} \geq V_{1\delta} - \varepsilon \qquad \forall \delta \in C.$$

π is called *optimal* if it is 0-optimal.

c) π is called $\bar{p}$-ε-optimal if

$$(20.3) \qquad V_\pi \geq V_\delta - \varepsilon \qquad \forall \delta \in C.$$

π is called $\bar{p}$-optimal, if it is $\bar{p}$-0-optimal.

d) Let W_1 be universally measurable.
π is called $(\gamma,p,\varepsilon)'$-optimal, if

$$(20.4) \qquad p(V_{1\pi} \geq W_1 - \varepsilon) \geq 1-\gamma.$$

Remarks. 1) The notion of (γ,p,ε)-optimality occurs in Hinderer (67); (p,ε)-optimality occurs in Blackwell (65); $\bar{p}$-ε-optimality occurs implicitely in Blackwell (67); $\bar{p}$-optimality occurs implicitely in Strauch (66) and in Hinderer (67) (under the name of M-p-optimality); $(p,\varepsilon)'$-optimality occurs in Strauch (66) (under the name of (p,ε)-optimality.

2) There are some elementary relations between these notions of optimality which are illustrated by fig.5 in which e.g. "$p \longrightarrow (p,\varepsilon)$" means: π is p-optimal iff it is (p,ε)-optimal for all $\varepsilon>0$, while e.g. "$p \dashrightarrow \bar{p}$" means: π is $\bar{p}$-optimal if it is p-optimal. Simple examples show that the indicated inclusions are proper. Without any assumptions on C only the existence of a randomized $\bar{p}$-ε-optimal plan (where $\varepsilon>0$ is arbitrary) can be guaranteed. Under mild conditions p-optimality coincides with $\bar{p}$-optimality (cf.theorem 20.1).

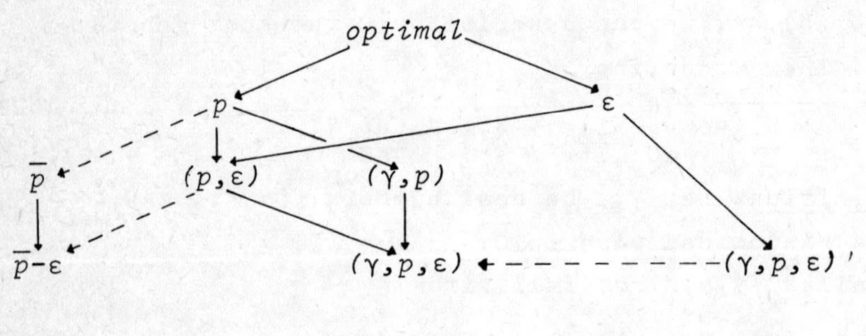

fig. 5

For non-stationary models notions of optimality have been used which take into account not only $V_{1\pi}$ but the whole sequence $(V_{n\pi})$. We shall use the

<u>Definition</u>. (Hinderer (67)) Let $p=(p_n)$ be a sequence of probability measures on $\mathcal{G}_n$, let $\varepsilon=(\varepsilon_n)$ be a sequence of real numbers $\varepsilon_n \geq 0$. The plan $\pi \in C$ is called *strongly* (p,ε)-*optimal* or *strongly* ε-*optimal*, if

$$p_n(V_{n\pi} \geq V_{n\delta} - \varepsilon_n) = 1, \quad n \in \mathbb{N}, \ \delta \in C,$$

or

$$V_{n\pi} \geq V_{n\delta} - \varepsilon_n, \quad n \in \mathbb{N}, \ \delta \in C,$$

respectively. The plan π is called *strongly optimal*, if it is ε-optimal for $\varepsilon=(0,0,\ldots)$.

Dynkin (65) calls strongly ε-optimal plans just ε-optimal, while Sirjaev (67) calls them uniformly ε-optimal. The works of Dynkin (65), Hinderer (67) and Sirjaev (67) show that strongly optimal plans play in non-stationary models the same role as optimal plans in stationary models. In our opinion, the notion of a strongly optimal plan is less appropriate for the description of practical problems than the notion of $\bar{p}$-optimality.

Another widely used notion of optimality is that using an average cost criterion. Here one can weaken the hypothesis that case (EN) or case (EP) holds to the assumption that either all reward functions are bounded from above or all are bounded from below (cf. Derman (62)).

Let us turn to the *average cost* criteria. Put $W_{n\pi}:=E_\pi[r_n o(\chi_n,\alpha_n)|\zeta_1=\cdot]$.

1) Let the model be stationary, and assume S and A finite and $\beta=1$. Assume that for any stationary plan $f=(\varphi,\varphi,\ldots)$ the recurrent states of the stochastic matrix $p(\varphi):=(p(i,\varphi(i),j)$ are aperiodic.
Then there exists $g_f:=\lim_n W_{nf}$, and the *stationary* plan f^* is called *average-optimal* (Howard (60/65)), if $g_{f^*}=\sup_f g_f$. It should be noted that one optimizes here only within the class of stationary plans.

2) Let the model be as in 1) with the exception that no assumption on $p(\varphi)$ is made. Then there exists $g_f:=\lim_n \frac{1}{n}\sum_1^n W_{\nu f}$, and f^* is called *average-optimal*, if $g_{f^*}=\sup_f g_f$.

3) Let the model be as in 2). If π is not stationary, then $\lim_n \frac{1}{n}\sum_1^n W_{\nu f}$ need not exist. It has become customary (cf. e.g. Derman (62)) to use then $g_\pi:=\underline{\lim}_n \frac{1}{n}\sum_1^n W_{\nu f}$, obviously for reasons similar to the min-max principle. (As far as we know, only Ogawara (64) states this plausible motivation.)

4) Put $V_\pi^n:=\sum_1^n W_{\nu\pi}$, the expected reward for the first n time periods. Brown (65) calls π^* *optimal*, if $V_{\pi^*}^n \geq V_\pi^n$ for all π.

5) Lippman (69) calls π^* *overtaking optimal*, if

$$\underline{\lim}_n (V_{\pi^*}^n - V_\pi^n) \geq 0 \quad \text{for all } \pi.$$

This notion of optimality is particularly attractive from an intuitive point of view.

6) Lippman (69) calls π^* *average return optimal*, if

$$\underline{\lim}_n \frac{1}{n}(V_{\pi^*}^n - V_\pi^n) \geq 0 \text{ for all } \pi.$$

7) Lippman (69) calls π^* *average overtaking optimal*, if

$$\underline{\lim}_n \frac{1}{n}\sum_1^n (V_{\pi^*}^\nu - V_\pi^\nu) \geq 0 \text{ for all } \pi.$$

That notion of optimality has been introduced by Veinott (66) and has been used by Denardo and Miller (68) under the name of *Veinott optimality*.

The average cost criteria 3)-7) make sense also in aritrary non-stationary models, provided $V_{n\pi}$ exists for any n and π. As shown in the given references, there are many relations between these notions of optimality (cf.also part C of the present section).

Finally we mention another notion of optimality, introduced by Kall (64) and studied by Hinderer (67).

Let $r\in\mathbb{N}+\{\infty\}$. Then π^* is called *<r>-optimal*, if, roughly spoken, one takes at any time n and any history h that action that is the initial action of a plan that is optimal for the next r steps. More precisely, π^* is <r>-optimal, if there exists for any $n\in\mathbb{N}$ a plan $\delta^{(n)}$ such that

$$\Lambda_{n\pi^*}E_{\delta(n)}\left[\sum_{\nu=n+1}^{n+r-1} r_\nu o(\chi_\nu,\alpha_\nu)\,|\,\psi_{n+1} = \cdot\right]$$

$$= \sup_\pi E_\pi\left[\sum_{\nu}^{n+r-1} r_\nu o(\chi_\nu,\alpha_\nu)\,|\,\psi_n = \cdot\right] , \; n\in\mathbb{N}.$$

It has been shown by Hinderer (67) that π^* is <∞>-optimal, iff it is strongly optimal.

For stationary models there are in the discounted case other notions of optimality in use. We fix some bounded measurable reward function r and use different discount rates $\beta\in(0,1\rangle$. Denote V_1 and $V_{1\pi}$ in the model with discount rate β by V_1^β and $V_{1\pi}^\beta$. Let us call π *β-optimal*, if it is optimal in the DM with discount rate β.

a) Blackwell (62) calls a plan π *optimal*, if there exists a $\beta_o\in(0,1)$ such that π is β-optimal $\forall\beta\in(\beta_o,1)$. In order to avoid confusion, we shall call such a plan *∞-optimal*.

b) Blackwell (62) calls a plan π *nearly optimal* (1-optimal in the terminology of Veinott (66)), if

$$\lim_{\beta\uparrow 1} (V_1^\beta - V_{1\pi}^\beta) = 1.$$

c) Put $\rho:=(1-\beta)/\beta$. Veinott (69) calls a plan π *n-optimal* ($n\in\{-1,0,1,...\}$) if

$$\underline{\lim}_{\rho\downarrow 0}\; \rho^{-n}[V_{1\pi}^\beta - V_{1\sigma}^\beta] \geq 0 \;, \; \delta\in C.$$

d) Maitra (65) mentions two other criteria, weaker than ∞-optimality.

B. Some results for general sets of admissible plans.

Without further assumptions on C one cannot hope to get reasonable results, as the following example shows.

Example 1. (There need not exist (γ,p,ε)-optimal plans.) Consider a Markovian model with
$S:=A:=\{0,1\}$, $D_n(s)=A$, $p_1(s,a,j):=\delta_{sj}$ (i.e. the system remains in the initial state), $p(0)=p(1)=1/2$, $r_n(s,a):=2^{-n}\delta_{sa}$, $C:=\{\pi^o,\pi^1\}$, where $\pi_n^i(h,\{j\})=\delta_{ij}$, i.e. π^i selects always i. It is easily seen that $V_{1\pi^i}(j)=\delta_{ij}$. Hence, for $\gamma\in(0,2^{-1})$ and $\varepsilon\in(0,1)$, neither π^o nor π^1 is (γ,p,ε)-optimal, though both are $\bar{p}$-optimal.

Now we are going to define properties of C which are suggested by results of Blackwell (65) and which exclude "pathological" models as in example 1. We remind of the notion of a plan, generated by a countable set $(\delta^i,i\in\mathbb{N})$ of plans, introduced in section 19. In addition, if there are sets $B_i\in\mathfrak{T}$ such that $\pi_n(h,\cdot)=\delta_n^i(h,\cdot)$ whenever $s_1\in B_i$, we shall say that π is generated by (δ^i) and (B_i).

Definition. The set C of plans is called *stable* (σ-*stable*), if every plan, which is generated by a finite (countable) set of plans in C belongs to C.

Of course, C is stable if every plan, which is generated by two plans in C, belongs to C.

The set Δ is an example of a σ-stable set of plans.

Theorem 20.1. If C is stable and W is finite, then the plan $\pi\in C$ is $\bar{p}$-optimal iff it is p-optimal.

Proof. Let π be $\bar{p}$-optimal, and assume that there exists a plan $\delta\in C$ such that $p(V_{1\delta}>V_{1\pi})>0$, hence $p(B):=p(V_{1\delta}\geq V_{1\pi}+\alpha)=\beta$ for some $\alpha>0,\beta>0$, Let σ be the plan generated by (δ,π) and $(B,S-B)$. We have $pV_{1\sigma}\leq pV_{1\pi}$ since $\sigma\in C$. On the other hand, $pV_{1\sigma}=p(V_{1\delta}\cdot 1_B+V_{1\pi}\cdot 1_{S-B})\geq pV_{1\pi}+\alpha\beta$. But this is a contradiction, as W is finite. ⌋

An easy modification of example 2 below shows, that theorem 20.1 is false without finiteness of W.

Lemma 20.2. If C is stable, there exists a countable set $M\subset C$ such that $p(V_{1\pi}\leq\sup_{\delta\in M} V_{1\delta})=1,\pi\in C$.

Proof. If $V_{1\pi}$ is bounded for all π in C, Lemma 20.2 follows easily from the proof of a corresponding theorem of Blackwell (65), (theorem 1). For the general case we use a kind of truncation. α) Assume case (EN). For $m \in \mathbb{N}$ and every function $f: S \to \overline{\mathbb{R}}$ we put $f^m := \max(f,-m)$. For $\pi \in C$ we get $-m \leq pV^m_{1\pi} \leq pV^+_{1\pi} \leq \sum \| r^+_i \|$. Therefore

$$(20.4) \qquad -\infty < \sup_{\pi} pV^m_{1\pi} < \infty, \quad m \in \mathbb{N}.$$

β) For fixed m there exists a countable set $M_m \subset C$ such that $\sup_{M_m} pV^m_{1\delta} = \sup_{C} pV^m_{1\pi}$. For $g := \sup_m \sup_{M_m} V_{1\delta}$ we assert:

$$(20.5) \qquad p(V^m_{1\pi} \leq g^m) = 1, \quad \pi \in C.$$

Assume (20.5) to be false. Then $p(B) := p(V^m_{1\tau} \geq g^m + \alpha) = \beta$ for some $\tau \in C$ and some $\alpha > 0, \beta > 0$. For $\delta \in M_m$ we denote the plan, generated by (τ,δ) and $(B, S-B)$, by δ'. Since δ' belongs to C, we have

$$(20.6) \qquad \sup_{M_m} pV^m_{1\delta'} \leq \sup_{C} pV^m_{1\pi} = \sup_{M_m} pV^m_{1\delta}.$$

On the other hand, according to the definition of g and B, we get $pV^m_{1\delta'} \geq \int_B (V^m_{1\delta} + \alpha)\,dp + \int_{S-B} V^m_{1\delta}\,dp \geq pV^m_{1\delta} + \alpha\beta$ for all $\delta \in C$.

This implies, by (20.4), that $\sup_{M_m} pV^m_{1\delta'} > \sup_{M_m} pV^m_{1\delta}$. But this contradicts (20.6), therefore (20.5) is verified.

γ) From (20.5) we get $p(-m \leq V_{1\pi} \leq g) = p(-m \leq V_{1\pi} \leq g; V^m_{1\pi} \leq g^m) = p(-m \leq V_{1\pi}; V^m_{1\pi} \leq g^m) = p(-m \leq V_{1\pi})$, which implies

$$p(V_{1\pi} \leq g) = p(V_{1\pi} = -\infty) + \lim_m p(-m \leq V_{1\pi} \leq g)$$
$$= p(V_{1\pi} = -\infty) + \lim_m p(-m \leq V_{1\pi}) = 1.$$

This proves the theorem for case (EN). For case (EP) the proof is similar, using $f^m := \min(f,m)$. ┘

From lemma 20.2 we get, using an idea in Blackwell (65), the following result.

<u>Theorem 20.3</u>. (Hinderer (67)). Let ε, γ be real numbers, $\varepsilon > 0, 0 < \gamma < 1$.

a) If C is stable, there exists a (γ, p, ε)-optimal plan.

b) If C is σ-stable, there exists a (p,ε)-optimal plan.

Proof. Denote the set M of lemma 20.2 by $M = \{\delta^1, \delta^2, \ldots\}$, and put $g := \sup_n V_{1\delta^n}$. Then the sets $B_n := [V_{1\delta^\nu} < g - \varepsilon, 1 \leq \nu < n; V_{1\delta^n} \geq g - \varepsilon]$, $n \in \mathbb{N}$, form a measurable partition of S_1. Choose m so large that $p(\sum_1^m B_n) \geq 1-\gamma$ holds. The plan σ, which is generated by $(\delta^1,\delta^2,\ldots,\delta^{m+1})$ and $(B_1,B_2,\ldots,B_m,\sum_{m+1}^{\infty} B_n)$ belongs to C. We have $p(V_{1\sigma} \geq g-\varepsilon) \geq 1-\gamma$, and the assertion follows from lemma 1.

b) The proof of part a) runs through with $m=\infty$ and $\gamma=0$.⌋

Examples of Blackwell ((65), example 1 and 2) show, that neither p-optimal nor ε-optimal plans need exist if C is σ-stable.

Example 2. (C is stable, but there is no (p,ε)-optimal plan.) We assume a Markovian model with $S:=\mathbb{N}$; $A:=\{0,1\}$; $D_n(h)=A$; $p(s,a,j):=\delta_{sj}$, i.e. the system remains in the initial state; p any probability with unbounded support; $r_n(s,a):=2^{-n}\delta_{a1}$. We have case (C), therefore every plan is admissible. We take C to be the set of plans π, such that π is generated by a finite subset (depending on π) of $(\pi^i, i\in\mathbb{N})$, where $\pi_n^i(h,1):=\delta_{is_1}$. It is easily seen, that every such π has the form

$$(20.7) \qquad \pi_n(h,1):=\sum_{i\in I} 1_{B_{ni}}\cdot\delta_{is_1} ,$$

for some finite $I\subset\mathbb{N}$ and some $B_{ni}\subset H_n$. From (20.7) we conclude that C is stable and that $(\pi_n r_n)(h)= =2^{-n}\pi_n(h,1)\leq 1_I(s_1)$. One easily verifies that

$$(20.8) \qquad V_{1\pi} \leq 1_I .$$

Let m be an integer not in I, and define $\delta\in C$ by

$$\delta_n(h,\cdot):=\begin{cases}\pi_n^i(h,\cdot), & s_1=i,\ n\in I,\\ \pi_n^m(h,\cdot), & \text{otherwise.}\end{cases}$$

Obviously δ belongs to C, and we have $\delta_n(h,1)=1_J(s_1)$, where $J:=I+\{m\}$. Hence $V_{1\delta}(s_1)=1_J(s_1)$, which implies $V_{1\delta}(m)=1$ and $V_{1\pi}(m)=0$. Since π is arbitrary, there does not exist a (p,ε)-optimal plan for any $\varepsilon\in(0,1)$.⌋

Example 3. (If C is arbitrary, (G_n) need not satisfy the optimality equation.) Consider the Markovian model of example 1 and take $C:=\{\pi^0,\pi^1\}$, where $\pi_n^i(h,j):=\delta_{ij}$ for $n>1$ and $\pi_1^i(s,j):=1/2$. One easily checks that

$G_{2\pi i}(s,a,j)=\frac{1}{2}\cdot\delta_{ij}$ and $G_{1\pi i}(1)=(1+2\delta_{i1})/4$, hence $G_2\equiv 1/2$ and $\sup_{a\in A} L_1G_2(1,a)=1$. On the other hand we have $G_1(1)=\frac{3}{4}$.

C. A short summary of results of stochastic dynamic programming, not treated in the present work.

We give a selection of references and some comments to several problems of dynamic programming. This survey is by no means complete. More references may be found in Menges (68), Sirjaev (67) and Weinert (68).

To Section 2. The models investigated in the main sources of the present work are contained in our general model in the following way: Blackwell (65), Strauch (66) and Maitra (68) use stationary models (cf.sections 6 and 18), where in addition D(s) equals A. Dynkin (65) and Sirjaev (67) use non-stationary models with incomplete information (cf. section 7). After reducing their models to models in the sense of our definition one realizes that Sirjaev's models coincide with ours (apart from different assumptions on r_n), whereas in Dynkin (65) S and A are countable, $p_n(h,a,j)$ and $r_n(h,a)$ do not depend on $a_1,a_2,\ldots,a_{n-1}$, and only deterministic plans are admitted. (Dynkin and Sirjaev use a terminology where the time index n does not appear explicitely; the resulting formulae are very elegant, but for the non-expert the notation seems to be less comprehensible than the formally more involved one used in the present work.) Furukawa investigated models where only some components are non-stationary; in Furukawa (67) and Furukawa (68) the model is stationary except that there may be a dependence on the time parameter n of the discount factor or of the Markovian transition law, rexpectively. Furakawa generalizes results of Blackwell (65) to his models. It is shown in Hinderer (67) that a partial generalization is possible also for our general model. We did not completely explore the possibility of generalizing the results of Blackwell (65) and Strauch (66), since we used only the notion of $\bar{p}$-optimality. Consequently, some of our results on the existence of a $\bar{p}$-optimal plan can be sharpened to a result on the existence of an optimal plan.

To Section 3. Bellman (57), p.115 contains historical notes on the history of the principle of optimality.

To Section 5. Theorem 5.3 is that sufficient criterion for the existence of an optimal plan that is widely used in the literature.

To Section 6. The case of finite state and action space has been treated in Blackwell (62), Ogawara (64) and Krylov (65), the case of a countable state space and a finite action space is dealt with in Maitra (65). The construction of optimal plans by means of Howard's method of policy-iteration is studied by Howard (60/65); additional information on that problem is contained in Blackwell (62), Brown (65) and Veinott (66).

To Section 8. There are many interesting problems in *deterministic* dynamic programming. We do not give any particular reference but refer to the books cited in section 1.

To Section 10. A rigorous presentation of Bayesian decision models for general state and action space has been given by Rieder (70). Another paper on dynamic programming under uncertainty is that of Miyasawa (68).

To Section 12. Readers interested in the measure-theoretic problems of decision models should study the stimulating work of Dubins and Savage (65) who circumvent the measurability problems to some extent by using finitely-additive measures.

To Section 13. Maitra (69) shows that under certain topological assumptions on A, q and r in a stationary model with general state and action space the function V_1 is measurable.

To Section 14. The OE (14.4) is stated in its full generality in Sirjaev (67). However, the proof given there assumes implicitely that V_n is measurable and that there exists for any $\varepsilon>0$ a strictly ε-optimal plan (which indeed implies measurability, cf.theorem 19.3).

To Section 20.

Average cost criteria.

α) Howard (60/65) shows by means of his method of policy iteration the existence of an average optimal plan (within the set of stationary plans) under the assumptions stated under 1). Manne (60) and Wolfe and Dantzig (62) show how the problem may be treated by linear programming. The possibility to restrict attention in Howards problem to stationary plans has been shown by Wagner (60) (using linear programming) by Derman (62) and Viskov and Sirjaev (64) (using discounted models). The case of a compact set D(s) is treated by Martin-Löf (67) by means of convex programming.

β) The extension of problem α) to the case of a countable state space and finite action space has been studied by Derman (66) and Ross (68) who show that an average optimal stationary plan exists only under restrictive assumptions. Derman (66) also extends Howards policy-iteration method to the denumerable state case.

γ) The extension of problem α) to the case of a general state space and finite action space has been considered by Taylor (65) and Ross (68a).

δ) Denardo and Miller (68) verify the conjecture of Veinott that there exists in a stationary model with finite state space and finite sets D(s) a stationary average overtaking plan. Lippman (69) proves that for the same model a plan is average overtaking optimal iff it is 1-optimal.

Small interest rate.

α) Miller and Veinott (69) give a finite algorithm for the construction of a stationary ∞-optimal plan.

β) An important paper on small interest rates is that of Veinott (69).

Extension to Semi-Markov processes.

Here we mention the papers of Jewell (63) and (63a), de Cani (64), Howard (64) and Osaki and Mine (68).

Extension to the case of a continuous time parameter.

A few references are de Leve (64), Åström (65), Rykov (66), Sirjaev (67), Miller (68) and Miller (68a).

Appendix 1. List of symbols and conventions.

$a: = b$	a is by definition equal to b
$R(a),\ a\in A$	relation $R(\cdot)$ holds for all a in A
$\mathbb{N}$	set of positive integers
$\mathbb{N}_o$	$\mathbb{N} + \{0\}$
$(\mathbb{R},\mathcal{B}_1)$	set of real numbers with σ-algebra $\mathcal{B}_1$ of Borel sets
$\mathbb{R}^d$	d-dimensional euclidean space
$\overline{\mathbb{R}}$	set of extended real numbers
$a^{\pm}:=\max(\pm a,o)$	the positive and negative part of $a\in\overline{\mathbb{R}}$
(a_n)	sequence of elements $a_1,a_2,\ldots$ or $a_o,a_1,\ldots$
$\lrcorner$	end of proof
1_A	indicatorfunction, defined by $1_A(x): = \{^0_1$ if $x \{^{\notin A}_{\in A}$
$\sigma(\mathcal{L})$	the σ-algebra generated by the system $\mathcal{L}$ of subsets of some set Ω:= smallest σ-algebra containing $\mathcal{L}$
$\mathcal{F}_1\otimes\mathcal{F}_2$	product-σ-algebra, determined by the component σ-algebras $\mathcal{F}_1$ and $\mathcal{F}_2$
$\mathcal{P}(A)$	system of all subsets of the set A
S^n	n-fold cartesian product of the set S
$B\cap\mathcal{L}$	$\{BC:\ C\in\mathcal{L}\}$, the trace of the system $\mathcal{L}$ on the set B.
iff	if and only if
$pr(C,A)$	the image of $C\subset A\times B$ under the projection from $A\times B$ into A
$pr_1:A_1\times A_2\to A_1$	projection map
A^c	the complement of the set A

In sequences, series, sup, $\underline{\lim}$, lim etc. the indices run through $\mathbb{N}$ if not otherwise indicated. A set is called countable, if it is either finite or infinitely denumerable. If a countable set A needs a σ-algebra, we always use the system of all subsets of A. For any function $f:A\times B\to C$ and any fixed a in A we denote by $f(a,\cdot)$ that function $g:B\to C$ which assumes the value $f(a,b)$ at point b. A counting density (density with respect to counting measure) on a countable set A is a function $p:A\to\mathbb{R}$ such that $p\geq 0$ and

$\sum_{a \in A} p(a)=1$. (The usual term "probability function" has no generally accepted German translation.) If $f:A \to B$ and $g:B \to C$ are maps then $g \circ f$ is the map from A to C whose value at a is $g(f(a))$. A map $f:A \to B$ is called surjective if it is a map onto and bijective if it is a map onto and one-to-one. If A is a subset of $B \times C$, and if b belongs to B, then $A_b := \{c \in C : (b,c) \in A\}$ is called the section of A at b.
δ_{ij} is the Kronecker symbol: $\delta_{ij} = \{ {0 \atop 1}$ for $i {\neq j \atop = j}$.
A maximum point of a function $g:B \to \overline{\mathbb{R}}$, bounded from above, is a point $x \in B$ at which f attains its supremum.
A singleton is a set consisting of one point.

Appendix 2. Some notions and auxiliary results from probability theory.

In this section we collect some notions and auxiliary results for the mathematically less experienced reader.

A. The system $\overline{\mathbb{R}}$ of extended real valued numbers.

In many mathematical investigations one encounters sequences (x_n) of real numbers with the property that x_n is larger than any given real number y as soon as n is larger than some number $n_o(y)$. It is a generally adopted custom to describe that property of (x_n) by saying that (x_n) converges to infinity if n converges to infinity. In the theory of integration which is basic for the calculus of expectations often situations similar to that one described above occur. Hence there arose the need for a rigorous formalization of the concept of "infinity". This is done by adjoining to the set $\mathbb{R}$ of real numbers two arbitrary elements (not already contained in $\mathbb{R}$), that are usually denoted by $-\infty$ and $+\infty$, and to extend *by definition* the arithmetical operations and some structures in $\mathbb{R}$ to the set $\overline{\mathbb{R}}$ in such a way that one gets a useful instrument for describing situations similar to that one cited above.

In the following x,y,z,... are real numbers and a,b,c,... are *extended real numbers*, i.e. elements of $\overline{\mathbb{R}}$.

1) $\overline{\mathbb{R}}$ is totally ordered by the usual ordering in $\mathbb{R}$ and by the definition:

$$-\infty < x \quad , x \in \mathbb{R} ,$$
$$x < \infty \quad , x \in \mathbb{R} ,$$
$$-\infty < \infty$$

2) The operation 'addition' is extended by defining:

$$x+(\pm\infty) := (\pm\infty)+x := \pm\infty, \quad x \in \mathbb{R}$$
$$(\pm\infty)+(\pm\infty) := \pm\infty .$$

The terms $(\pm\infty)+(\mp\infty)$ remain undefined. It follows that a finite series $\sum_1^n a_\nu$ of extended real numbers is defined iff the sequence $(a_1,\dots,a_n)$ contains at most one of the numbers $\pm\infty$, or equivalently, iff $\sum_1^n a_\nu^+ < \infty$ or $\sum_1^n a_\nu^- < \infty$. The value of $\sum_1^n a_\nu$ is independent of the arrangement of the terms and equals $\sum_1^n a_\nu^+ - \sum_1^n a_\nu^-$.

3) The operation 'subtraction' is extended by $a-b:=a+(-b)$,

whenever $a+(-b)$ is defined.

4) The operation 'multiplication' is extended by defining

$$a\cdot(\pm\infty):=(\pm\infty)\cdot a:=\begin{cases}\pm\infty \\ 0 \\ \mp\infty\end{cases}, \text{ for } a \begin{cases} >0 \\ =0 \\ <0\end{cases}, \quad a\in\overline{\mathbb{R}}$$

The definition $0.(\pm\infty):=0$ seems to be rather artificial but it is very useful in measure theory.

5) The operation 'division' is extended by defining

$$\frac{1}{\pm\infty}:=0, \ \frac{1}{0}:=\infty, \ \frac{a}{b}:=a\cdot\frac{1}{b}, \quad a,b\in\overline{\mathbb{R}}.$$

The definition $\frac{a}{b}:=a\cdot\frac{1}{b}$, which in particular implies $\frac{0}{0}:=\frac{\pm\infty}{\pm\infty}:=\frac{\mp\infty}{\pm\infty}:=0$, is not standard, but very convenient when dealing with conditional probabilities and conditional densities.

B. Expectations of discrete distributions.

When dealing with expectations of discrete probability distributions it is useful to admit that the expectation attains the value $+\infty$ or $-\infty$. Therefore one has to study the convergence of series of extended real valued numbers.
If (a_n) is a sequence of numbers $a_n\in\overline{\mathbb{R}}$, then the definition of $\sup a_n$, $\underline{\lim}\ a_n$, $\lim_n a_n$ etc. as elements of $\overline{\mathbb{R}}$ is obvious.
If $\lim_n a_n\in\overline{\mathbb{R}}$ exists we shall say that the sequence (a_n) converges in $\overline{\mathbb{R}}$.
Let (a_n) be a sequence of numbers in $\overline{\mathbb{R}}$ that contains at most one of the numbers $+\infty,-\infty$. Then $\sum_1^n a_\nu$ is well-defined for any $n\in\mathbb{N}$. We shall say that $\sum a_\nu$ converges (in $\overline{\mathbb{R}}$) to the value $a\in\overline{\mathbb{R}}$ iff $\lim_n \sum_1^n a_\nu$ exists and equals a. We shall need the simple

<u>Lemma A1</u>. If $\sum a_n$ converges in $\overline{\mathbb{R}}$, then $(\sum a_n)^{\pm}\leq\sum a_n^{\pm}$.

<u>Proof</u>. It is easily seen that $(a+b)^+\leq a^++b^+$ for any $a,b\in\overline{\mathbb{R}}$ for which $a+b$ is defined. By induction the same relation is proved for any finite sum. The map $x\to x^+$ is clearly continuous in $\overline{\mathbb{R}}$, hence $(\sum a_n)^{\pm}=\lim_N(\sum_1^N a_n)^{\pm}\leq\overline{\lim_N}\sum_1^N a_n^{\pm}=\sum a_n^{\pm}$. ⌋

Let $(q(i),i\in I)$ be a counting density on a denumerable set I, and let $X:I\to\overline{\mathbb{R}}$ be a random variable on I (i.e. an arbitrary function) such that at most one of the values $\pm\infty$ occurs

in the sequence $(X(i)q(i))$. The definition of the expectation of X by means of $EX := \sum_{i I} X(i)q(i)$ is clear if I is finite. It is meaningless if I is infinite denumerable unless we have arranged the terms $X(i)q(i)$ in some way by means of a bijection $g: \mathbb{N} \to I$, which would allow to define $EX := \sum_n X(g(n))q(g(n))$. In general, e.g. if I is the set of lattice points in euclidean n-space, there does not exist a natural arrangement of the terms $X(i)q(i)$. Therefore one agrees to define EX at most in those cases in which $\sum_n X(g(n))q(g(n))$ converges (in $\overline{\mathbb{R}}$) for any bijection $g: \mathbb{N} \to I$.
Let us call a formal series $\sum_{i \in I} a_i$, where I is infinitely denumerable and where at most one of the numbers $+\infty, -\infty$ occurs among the a_i's, *unconditionally convergent* if $\sum_n a_{g(n)}$ converges in $\overline{\mathbb{R}}$ for any bijection $g: \mathbb{N} \to I$. One easily proves:

(i) $\sum a_i$ converges unconditionally if $a_i \geq 0 \quad i \in I$.

(ii) If $\sum_i a_i$ converges unconditionally, then $b := \sum_n a_{g(n)}$ is independent of g, and b is called the value of $\sum a_i$.

(iii) $\sum a_i$ converges unconditionally iff either

α) $\sum a_i^+ < \infty$ or $\sum a_i^- < \infty$, or

β) $\sum a_i^+ = \sum a_i^- = \infty$ and $|a_{i_o}| = \infty$ for some i_o.

In case (α) we have $\sum a_i = \sum a_i^+ - \sum a_i^-$, in case (β) we have $\sum a_i = a_{i_o}$.
In particular, for sequences (a_i) of *real* numbers, case (α) characterizes unconditional convergence. Therefore it is customary to use the following

Definition. Let $(q(i), i \in I)$ be a counting density on a countable set I, and let $X: I \to \overline{\mathbb{R}}$ be a random variable, such that either $\sum X^+(i)q(i) < \infty$ or $\sum X^-(i)q(i) < \infty$. Then $EX := \sum_i X(i)q_i = \sum_i X^+(i)q_i - \sum_i X^-(i)q_i$ is called the expectation of X.

C. Calculus of expectations.

Let $(\Omega, \mathfrak{a}, P)$ be a probability space and $X: \Omega \to \overline{\mathbb{R}}$ be a random variable on the space. We shall use the usual notion of the expectation $EX := \int X dP$ as developped, e.g. in Loève (60). Let us mention some of its properties.

1) EX is defined if $X \geq 0$; then $0 \leq EX \leq \infty$.

2) EX is defined iff $EX^+ < \infty$ or $EX^- < \infty$; then $EX = EX^+ - EX^- \in \overline{\mathbb{R}}$, and X is called *quasi-integrable*.

X is called *integrable* if it is quasi-integrable and if $EX \in \mathbb{R}$ or equivalently, if $EX^+ < \infty$ *and* $EX^- < \infty$.

3) It is often important to know under what circumstances summation and expectation are interchangeable. Mostly used are the following two criteria, which are easily derived from the monotone convergence theorem.

Lemma A2. Let (X_n) be a sequence of extended real valued random variables.

a) If $X_n \geq 0$, then $E\sum X_n = \sum EX_n$.

b) If $\sum E|X_n| < \infty$, then $\sum EX_n$ converges in $\mathbb{R}$, $\sum X_n$ converges a.s. to an integrable random variable, and $E\sum X_n = \sum EX_n$.

We shall prove the following generalization of lemma A2.

Theorem A3. Let (X_n) be a sequence of extended real valued random variables such that $\sum EX_n^+ < \infty$ or $\sum EX_n^- < \infty$. Then we have:

(i) $\sum EX_n$ converges to $\sum EX_n^+ - \sum EX_n^-$.

(ii) $\sum X_n$ converges a.s. to a quasi-integrable random variable.

(iii) $E\sum X_n = \sum EX_n$.

(iv) $E(\sum X_n)^{\pm} \leq \sum EX_n^{\pm}$.

Proof. Let us assume that $\sum EX_n^+ < \infty$. (i) We have $EX_n^+ < \infty$, hence EX_n exists and $EX_n \leq EX_n^+ < \infty$ for every n. Hence there is defined $\sum_1^N EX_n = \sum_1^N EX_n^+ - \sum_1^N EX_n^-$ for any N, which converges in $\overline{\mathbb{R}}$ to $\sum EX_n^+ - \sum EX_n^-$ for $N \to \infty$. (ii) From Lemma A2 we get $E\sum X_n^+ = \sum EX_n^+ < \infty$, hence $\sum X_n^+ < \infty$ a.s. . Therefore there is defined a.s. the random variable $\sum_1^N X_n = \sum_1^N X_n^+ - \sum_1^N X_n^-$, which converges a.s. to the random variable $1_A \cdot \sum X_n^+ - \sum X_n^-$, where $A := [\sum X_n^+ < \infty]$. In order to show that $\sum X_n$ is quasi-integrable we conclude from lemma A1 that $(\sum X_n)^+ \leq \sum X_n^+$ on A, which implies $E(\sum X_n)^+ = E(\sum X_n)^+ \cdot 1_A \leq E\sum X_n^+ = \sum EX_n^+ < \infty$. This shows that $\sum X_n$ is quasi-integrable and that statement (iv) holds. For part (iii) we use the general addition theorem for the expectation of the sum of two quasi-integrable random variables (cf.e.g.Neveu (65),p.41) which yields $E\sum X_n = E(\sum X_n^+ \cdot 1_A - \sum X_n^-) = E\sum X_n^+ \cdot 1_A - E\sum X_n^- = \sum EX_n^+ - \sum EX_n^-$, but the last term equals $\sum EX_n$, as shown above. ⌋

Remark. For *finite* sums the condition $\sum_1^N EX_n^+ < \infty$ or $\sum_1^N EX_n^- < \infty$ is not only sufficient but also necessary for the existence and equality of the terms $E\sum_1^N X_n$ and $\sum_1^N EX_n$.

Most important for the development of the theory in chapters 1 and 2 is the following immediate conclusion of theorem A3.

Theorem A4. Let (X_n) be a sequence of extended real valued random variables such that $\sum \|X_n^+\| < \infty$ or $\sum \|X_n^-\| < \infty$. Then the statements (i)-(iv) of theorem A3 hold.

Appendix 3. Conditional distributions and expectations.

A. *Elementary conditional probabilities* $P(A|B)$ are used in the usual sense: $P(A|B):=P(AB)/P(B)$, with the exception that we use this definition also when $P(B)=0$, in which case $P(A|B)$ equals zero according to the definition $\frac{0}{0}:=0$, made in appendix 2A. This convention has many advantages, e.g. the formula

$$P(\bigcap_1^n A_\nu) = P(A_1)P(A_2|A_1)\ldots P(A_n|\bigcap_1^{n-1} A_\nu)$$

is now valid without the assumption $P(\bigcap_1^{n-1} A_\nu)>0$.

B. *Transition probabilities* (cf.e.g.Loève (60), Neveu (65)).

Let $(\Omega_1,\mathfrak{A}_1)$ and $(\Omega_2,\mathfrak{A}_2)$ be measurable spaces. A map $q:\Omega_1\times\mathfrak{A}_2\to\overline{\mathbb{R}}$ is called a *transition probability* (or Markov kernel, or stochastic kernel) from Ω_1 into Ω_2 if

(i) $q(\omega_1,\cdot)$ is a probability measure on $\mathfrak{A}_2$,

(ii) $q(\cdot,B)$ is measurable, $B\in\mathfrak{A}_2$.

A transition probability may loosely be described as a family of probabilities which depends in a measurable way on some parameter ω_1. Transition probabilities are important tools for the construction of stochastic processes and find there an interpretation as conditional distributions in the way described below. In practical problems transition probabilities q are given by "transition densities", i.e. measurable non-negative maps $g:\Omega_1\times\Omega_2\to\mathbb{R}$, such that $q(\omega_1,B)=\int_B f(\omega_1,\omega_2)\nu(d\omega_2)$ holds for some σ-finite measure ν on $\mathfrak{A}_2$. The function f may often be interpreted as a conditional density.

Let v: $\Omega_1\times\Omega_2\to\overline{\mathbb{R}}$ be measurable and non-negative. Then $\omega_1\to\int q(\omega_1,d\omega_2)v(\omega_1,\omega_2)$ is a measurable map from Ω_1 into $\overline{\mathbb{R}}$ which we denote sometimes by qv. The following well-known theorem extends the Fubini theorem for probabilites.

<u>Theorem A5</u>. Let μ be a probability on $\mathfrak{A}_1$ and let q be a transition probability from Ω_1 into Ω_2. Then there exists a unique measure, denoted by μq or $\mu\otimes q$, on $\mathfrak{A}_1\otimes\mathfrak{A}_2$ such that

$\mu q(A_1 \times A_2) = \int_{A_1} \mu(d\omega_1) q(\omega_1, A_2)$, $A_1 \in \mathcal{A}_1, A_2 \in \mathcal{A}_2$. If $f: \Omega_1 \times \Omega_2 \to \overline{\mathbb{R}}$ is μq-quasi-integrable, then

$$\omega_1 \to \int q(\omega_1, d\omega_2) f(\omega_1, \omega_2)$$

is μ-a.s. defined as a μ-quasi-integrable function, and

$$\text{(A1)} \qquad \int f d(\mu q) = \int \mu(d\omega_1) \int q(\omega_1, d\omega_2) f(\omega_1, \omega_2).$$

In our short-hand notation for integrals, relation (A1) reads as

$$(\mu q) f = \mu(qf).$$

The extension of theorem A5 to the product of a finite number of measurable spaces is rather obvious. For countably many factors we have the following theorem of C.Ionescu Tulcea.

Theorem A6. Let $((\Omega_n, \mathcal{A}_n))$ be a sequence of measurable spaces. Let q_o be a probability on $\mathcal{A}_1$ and q_n a transition probability from $\times_1^n \Omega_\nu$ to Ω_{n+1}. Then there exists a unique probability ν (denoted by $q_o q_1 q_2 \ldots$ or $\bigotimes_o^\infty q_\nu$) on $\bigotimes_1^\infty \mathcal{A}_\nu$, such that

$$\nu((\times_1^n A_\nu) \times (\times_{n+1}^\infty \Omega_\nu)) = \int_{A_1} q_o(d\omega_1) \int_{A_2} q_1(\omega_1, d\omega_2) \ldots$$

$$\ldots \int_{A_n} q_{n-1}(\omega_{n-1}, d\omega_n), \; n \in \mathbb{N}, A_\nu \in \mathcal{A}_\nu.$$

Let $X_n : \times_1^\infty \Omega_\nu \to \Omega_n$ be the projection mapping. Then $\bigotimes_1^\infty q_\nu$ may also be characterized as the unique measure ν such that

$$\int f \circ (X_1, X_2, \ldots, X_n) d\nu = \int q_o(d\omega_1) \int q_1(\omega_1, d\omega_2) \ldots$$

$$\ldots \int q_{n-1}(\omega_{n-1}, d\omega_n) f(\omega_1, \ldots, \omega_n)$$

holds for any map $f: \times_1^n \Omega_\nu \to \overline{\mathbb{R}}$ for which $f \circ (X_1, \ldots, X_n)$ is ν-quasi-integrable.

For any fixed $\omega_1 \in \Omega_1$, the map $((\omega_2, \ldots, \omega_n), B) \to q_n((\omega_1, \omega_2, \ldots, \omega_n), B)$ is a transition probability from $\times_2^n \Omega_\nu$ into Ω_{n+1}. Hence there is defined the probability $(q_1 q_2 \ldots)(\omega_1, \cdot)$ on $\bigotimes_2^\infty \mathcal{A}_\nu$. One can prove that this probability depends measurably on ω_1 (cf. Neveu (65),p.162). It follows easily that $(q_o q_1 \ldots) = q_o \otimes (q_1 \ldots)$. More generally we have

Lemma A7. Under the assumptions of theorem A6 we have
$(q_o q_1 \ldots q_n) \otimes (q_{n+1} q_{n+2} \ldots) = (q_o q_1 q_2 \ldots)$, $n \in \mathbb{N}$.

Lemma A8. Let $(\Omega_n, \mathfrak{A}_n)$ $n \in \mathbb{N}$, be measurable spaces, let μ_o and ν_o be probabilities on $\mathfrak{A}_1$, let μ_n and ν_n be transition probabilities from $\overset{n}{\underset{1}{\times}} \Omega_i$ to Ω_{n+1}. If $\overset{\infty}{\underset{o}{\otimes}} \mu_n = \overset{\infty}{\underset{o}{\otimes}} \nu_n$, then also $\overset{\infty}{\underset{o}{\otimes}} \mu_n = \mu_o \nu_1 \mu_2 \nu_3 \ldots$.

Proof. We know that

$$\overset{n}{\underset{o}{\otimes}} \mu_i = \overset{n}{\underset{o}{\otimes}} \nu_i, \quad n \in \mathbb{N} \tag{A2}$$

holds, and have to show that

$$\overset{2m-1}{\underset{o}{\otimes}} \mu_i = \mu_o \nu_1 \mu_2 \ldots \nu_{2m-1} \tag{A3}$$

holds for $m \in \mathbb{N}$. We shall use induction on m. At first (A2) implies $\mu_o = \nu_o$ and $\mu_o \mu_1 = \nu_o \nu_1$, hence $\mu_o \mu_1 = \mu_o \nu_1$, which proves (A3) for m=1. Let us assume that (A3) is true for some $m \in \mathbb{N}$. Then we get from (A2) and the induction hypothesis
$\mu_o \mu_1 \ldots \mu_{2m+1} = (\nu_o \nu_1 \ldots \nu_{2m}) \nu_{2m+1} = (\mu_o \mu_1 \ldots \mu_{2m}) \nu_{2m+1} =$
$= (\mu_o \mu_1 \ldots \mu_{2m-1})(\mu_{2m} \nu_{2m+1}) = (\mu_o \nu_1 \ldots \nu_{2m-1})(\mu_{2m} \nu_{2m+1}) =$
$= \mu_o \nu_1 \ldots \nu_{2m+1}$. ┘

C. *Conditional distributions.*

Conditional expectations play an important role in dynamic programming since the terms accuring in the OE are of that type. The general notion of conditional expectations (via the Radon-Nikodym theorem) is not easily accessible to intuition. Fortunately, in many practical problems the conditional expectations may be defined as integrals with respect to transition probabilities. Since this is true for the optimization problems we are concerned with, we shall restrict ourselves completely to conditional expectations in the sense of the subsequent definition.

Definition. Let $(\Omega, \mathcal{F}, P)$ be a probability space, let $(\Omega_i, \mathfrak{A}_i)$, i=1,2, be measurable spaces, let $X_i : \Omega \to \Omega_i$ be measurable. Denote by P_{X_1} and $P_{(X_1,X_2)}$ the distribution of X_1 and (X_1, X_2), respectively. A *conditional distribution* of X_2 under the condition X_1 is a transition probability q from Ω_1 into Ω_2 such that $P_{(X_1,X_2)}=P_{X_1}\otimes q$. We shall then write $q\in P_{X_2|X_1}$.

<u>Remarks</u>. 1. There do not always exist conditional probabilities. This is one of the reasons why most of the results on decision models are not obtainable if the measurable structure on the state space and the action space are arbitrary. There exists a conditional distribution $q\in P_{X_2|X_1}$, if $(\Omega_2,\mathfrak{A}_2)$ is an euclidean n-space or more generally, if $(\Omega,\mathfrak{A}_2)$ is an SB-space (cf. theorem 12.4). 2. Conditional distributions, if they exist, are not unique. This is the reason for our notation $q\in P_{X_2|X_1}$. 3. The question of the existence and computation of conditional expectations is in practical problems often easily solved, since the probability $P_{(X_1,X_2)}$ is often defined by means of transition probabilities; e.g. the definition of μq in theorem A5 tells us immediately that $q\in(\mu q)_{pr_2|pr_1}$, where pr_i is the projection from $\Omega_1\times\Omega_2$ into Ω_i. Lemma A7 may be reformulated as

<u>Lemma A9</u>. Let $(\Omega_n,\mathfrak{A}_n)$, ν and X_n be as in theorem A6. Then $q_n\in\nu_{X_{n+1}|(X_1,X_2,\ldots,X_n)}$ and

$q_nq_{n+1}\ldots\in\nu_{(X_{n+1},X_{n+2},\ldots)|(X_1,\ldots,X_n)}$.

D. *Conditional expectations.*

<u>Definition</u>. Let $(\Omega,\mathfrak{F},P)$, $(\Omega_i,\mathfrak{A}_i)$ and X_i be as in the definition of a conditional distribution. If $f:\Omega_1\times\Omega_2\to\overline{\mathbb{R}}$ is measurable, then

$$E[f\circ(X_1,X_2)|X_1=\omega_1]:=\int q(\omega_1,d\omega_2)f(\omega_1,\omega_2)$$

is called the *conditional expectation* of $f\circ(X_1,X_2)$ under the condition $X_1=\omega_1$, whenever the integral exists.

<u>Remarks</u>. 1. The conditional expectation is unique, as soon as we have made a decision for an element $q\in P_{X_2|X_1}$. In all situations in chapters 1 and 2 it will be clear what

choice has been made. In particular, when we are dealing with the situation of theorem A6, we shall take q_n as conditional distribution of X_{n+1} under the condition $(X_1,...,X_n)$.

2. Properties of conditional expectations in the sense of the definition given above are directly derived from properties of ordinary expectations.

Literature

Books and papers are numbered by their year of appearance.

Abbreviations: AMS = Annals of Mathematical Statistics,
MS = Management Science,
TP = Theory of Probability and its Applications
TxPC = Transactions of the xth Prague Conference on Information Theory, Statistical Decision Functions, Random Processes,
ST = Selected Translations in Mathematical Statistics and Probability,
ZW = Zeitschrift für Wahrscheinlichkeitstheorie und verwandte Gebiete,
OR = Operations Research.

Aoki,M. (65) Optimal control of partially observable Markovian systems. J.Franklin Inst.280, 367-386.

Aoki,M. (67) Optimization of stochastic systems. Academic Press, New York.

Aris,R. (64) Discrete dynamic programming (An introduction to the optimization of staged processes). Blaisdell, New York.

Åström,K.J. (65) Optimal control of Markov processes with incomplete state information. (I) J.Math.Anal.Appl.10,174-205.

Bauer,H. (68) Wahrscheinlichkeitstheorie und Grundzüge der Maßtheorie. Walter de Gruyter, Berlin.

Beck,A. (64) On the linear search problem. Israel J.Math.2, 221-228.

Beck,A. (65) More on the linear search problem. Israel J.Math.3, 61-70.

Beckmann,M.J. (68) Dynamic Programming of Economic Decisions. Springer, Berlin-Heidelberg-New York.

Bellman,R. (57) Dynamic programming. Princeton Univ.Press, Princeton.

Bellman R. and S.Dreyfus (62) Applied dynamic programming. Princeton Univ.Press, Princeton.

Black,W.L. (65) Discrete sequential search. Information and Control 8, 159-162.

Blackwell,D. (62) Discrete dynamic programming. AMS 33, 719-726.

Blackwell,D. (62a) Notes on dynamic programming (unpublished). Dept. of Stat.,Univ. of Calif.

Blackwell,D. (65) Discounted dynamic programming. AMS 36, 226-235.

Blackwell,D. (67) Positive dynamic programming. Proc.of the 5th Berkeley Symp.1965, Vol.I, 415-418.

Blackwell,D. and C.Ryll-Nardzewski (63) Non-existence of everywhere proper conditional distributions. AMS 34, 223-225.

Boudarel,R; J.Delmas; P.Guichet (68) Commande optimale des processus. Tome 3: programmation dynamique et ses applications. Dunod, Paris.

Bourbaki,N. (58) Eléments de Mathématiques, Livre III: Topologie générale. Chap.9, Utilisation des nombres réels en topologie générale. Herman, Paris, 2e éd.

Breiman,L. (64) Stopping-rule problems. In: (Beckenbach,E.F.,ed.) Applied combinatorial mathematics. Wiley, New York, 284-319.

Brown,B.W. (65) On the iterative method of dynamic programming on a finite space discrete time Markov process. AMS 36, 1279-1285.

de Cani,J.S. (64) A dynamic programming algorithm for embedded Markov chains when the planning horizon is at infinity. MS 10, 716-733.

de Leve,G. (64) Generalized Markovian decision processes. Part I: Model and method. Part II: Probabilistic background. Mathematisch Zentrum, Amsterdam.

Denardo,E.V. and B.L.Miller (68) An optimality condition for discrete dynamic programming with no discounting. AMS 39, 1220-1227.

Derman,C. (62) On sequential decisions and Markov chains. MS 9, 16-24.

Derman,C. (66) Denumerable state Markovian decision processes - average cost criterion. AMS 37, 1545-1554.

Dieudonné,J. (60) Foundations of Modern Analysis. Academic Press, New York.

Dubins,L.; Freedman,D. (64) Measurable sets of measures. Pacific J.Math.14, 1211-1222.

Dubins,L.E. and L.J.Savage (65) — How to gamble if you must. McGraw-Hill, New York

Dynkin,E.B. (61) — Die Grundlagen der Theorie der Markoffschen Prozesse. Springer, Berlin-Göttingen-Heidelberg

Dynkin,E.B. (65) — Controlled random sequences. TP 10, 1-14

Furukawa,N. (67) — A Markov decision process with discounted rewards. Mem.Fa.Sci. Kyushu Univ.Ser.A 21, 241-248.

Furukawa,N. (68) — A Markov decision process with non-stationary transition laws. Bull.Math.Stat.13, 41-52.

Hadley,G. (64) — Nonlinear and dynamic programming. Addison-Wesley, Reading, Mass.

Hausdorff,F. (27) — Mengenlehre. Walter de Gruyter, Berlin.

Hinderer, K. (67) — Zur Theorie stochastischer Entscheidungsmodelle. Habilitationsschrift, Univ.Stuttgart. (The essential parts of Chapter I and II of that paper have been incorporated in the present treatise, whereas Chapter III, dealing with a specific applied problem, will be published elsewhere.)

Hinderer,K. (71) — Instationäre dynamische Optimierung bei schwachen Voraussetzungen über die Gewinnfunktionen. To be published in Abh.Math.Sem.Univ.Hamburg, Bd.35.

Howard,R.A. (60/65) — Dynamic programming and Markov processes. Technology Press and Wiley, New York. German revised translation by H.P.Künzi and P.Kall, Verlag Industr.Org., Zürich (1965).

Howard,R.A. (64) — Research in semi-Markovian decision structures. J.Operat.Res.Soc.Japan 6, 163-199.

Jacobs,O.L.R. (67) — An introduction to dynamic programming. (The theory of multistage decision processes.) Chapman and Hall, London.

Jewell,W.S. (63,63a) — Markov renewal programming. I.Formulation, finite return models. II.Infinite return models, example. OR 11, 938-948, 949-971.

Kall,P. (64) — Anwendung endlicher Markov-Ketten in der linearen Programmierung. ZW 3,89-109.

Karlin,S. (55) — The structure of dynamic programming models. Naval Res.Logist.Quart.2,285-294

Krylov, N.V. (65) Construction of an optimal strategy for a finite controlled chain. TP 10, 45-54.

Künzi,H.P.; Nievergelt,E.; Müller,O. (68) Einführungskurs in die dynamische Programmierung. Lecture notes in operations research and mathematical economics. Springer, Berlin.

Kuratowski,C. (48) Topologie I. Warszawa.

Kuratowski,C. (50) Topologie II. Warszawa.

Lippman,S.A. (69) Criterion equivalence in discrete dynamic programming. OR 17,920-923.

Loève,M. (60) Probability theory. Van Nostrand, Princeton, 2nd ed.

Mackey,G. (57) Borel structure in groups and their duals. Trans.Amer.Math.Soc.85, 134-165.

Maitra, A. (65) Dynamic programming for countable state systems. Sankhya 27A, 241-248.

Maitra,A. (68) Discounted dynamic programming on compact metric spaces. Sankhya 30A, 211-216.

Maitra,A. (69) A note on positive dynamic programming. AMS 40, 316-319.

Manne, A.S. (60) Linear programming and sequential decisions. MS 6, 259-267.

Martin,J.J. (67) Bayesian decision problems and Markov chains. Wiley, New York.

Martin-Löf,A. (67) Existence of a stationary control for a Markov chain maximizing the average reward. OR 15, 866-871.

Matula,D. (64) A periodic optimal search. Amer.Math. Monthly 71, 15-21.

Menges,G. (68) Bibliographie zur statistischen Entscheidungstheorie 1950-1967. Westdeutscher Verlag, Köln.

Miller,B.L. (68) Finite state continuous time Markov decision processes with an infinite planning horizon. J.Math.Anal.Appl. 22, 552-569.

Miller,B.L. (68a) Finite state continuous time Markov decision processes with a finite planning horizon. SIAM J.Control 6, 266-280.

Miller,B.L. and A.F.Veinott (69) Discrete dynamic programming with a small interest rate. AMS 40, 366-370.

Miyasawa, K. (68) Information structures in stochastic programming problems. MS 14,275-291.

Natanson,I.P. (61) Theorie der Funktionen einer reellen Veränderlichen. Akademie-Verlag, Berlin.

Nemhauser,G.L. (66) Introduction to dynamic programming. Wiley, New York.

Neuman,K. (69) Dynamische Optimierung. Bibliographisches Institut, Mannheim.

Neveu,J. (65) Mathematical foundations of the calculus of probability. Holden-Day, San Francisco.

Ogawara,M. (64) A note on discrete Markovian decision processes.Bull.Math.Statist. 11, 35-42.

Osaki,S. and H.Mine (68) Linear programming algorithms for semi-Markovian decision processes. J.Math.Anal.Appl.22, 356-381.

Radner,R. (67) Dynamic programming of economic growth. Activity analysis in the theory of growth and planing. Macmillan, London.

Rieder,U. (70) Bayessche und verwandte dynamische Entscheidungsmodelle. Diplomarbeit Inst.Math.Stochastik, Univ.Hamburg.

Ross,S. M. (68) Non-discounted denumerable Markovian decision models. AMS 39, 412-424.

Ross,S.M. (68a) Arbitrary state Markovian decision processes. AMS 39, 2118-2122.

Rykov,V.V. (66) Markov decision processes with finite state and decision spaces. TP 11, 302-310.

Sawaragi,Y. and T.Yoshikawa (70) Discrete-time Markovian decision processes with incomplete state observation. AMS 41, 78-86.

Širjaev,A.N. (64) On the theory of decision functions and control of a process of observation based on incomplete information. T3PC (Prague 1962), 657-681. (Russian; English translation in ST 6 (1966),162-188.)

Širjaev,A.N. (67) Some new results in the theory of controlled random processes. T4PC (Prague 1965), 131-203. (Russian; English translation in ST 8, (1970), 49-130).

Strauch, R.E. (66) Negative dynamic programming. AMS 37, 871-890.

Sworder,D. (66) Optimal adaptive systems. Academic Press, New York.

Taylor,H.M., III (65) Markovian sequential replacement processes. AMS 36, 1677-1694.

Veinott,A.F. (66) On finding optimal policies in discrete dynamic programming with no discounting. AMS 37,1284-1294.

Veinott,A.F. (69) Discrete dynamic programming with sensitive optimality criteria. AMS 40, 1635-1660.

Viskov,O.V. and A.N.Širjaev (64) On controls leading to optimal stationary states. Trudy Mat.Inst. Steklov. 71,35-45, ST 6 (1966),71-83.

Wagner,H. (60) On the optimality of pure strategies. MS 6, 268-269.

Weinert,H. (68) Bibliographie über Optimierungsprobleme unter Ungewißheit. Operationsforschung und Mathematische Statistik 1, 137-151.

Wessels,J. (68) Decision rules in Markovian decision processes with incompletely known transition probabilities. Dissertation, TH Eindhoven.

White,D.J. (69) Dynamic programming. Oliver and Boyd, Edinburgh, London.

Wolfe,P. and G.B.Dantzig (62) Linear programming in a Markov chain. OR 10, 702-710.

Index of definitions and notations

Offsetdruck: Julius Beltz, Weinheim/Bergstr.

Lecture Notes in Operations Research and Mathematical Systems

Vol. 1: H. Bühlmann, H. Loeffel, E. Nievergelt, Einführung in die Theorie und Praxis der Entscheidung bei Unsicherheit. 2. Auflage, IV, 125 Seiten 4°. 1969. DM 12,– / US $ 3.30

Vol. 2: U. N. Bhat, A Study of the Queueing Systems M/G/1 and GI/M/1. VIII, 78 pages. 4°. 1968. DM 8,80 / US $ 2.50

Vol. 3: A. Strauss, An Introduction to Optimal Control Theory. VI, 153 pages. 4°. 1968. DM 14,– / US $ 3.90

Vol. 4: Einführung in die Methode Branch and Bound. Herausgegeben von F. Weinberg. VIII, 159 Seiten. 4°. 1968. DM 14,– / US $ 3.90

Vol. 5: L. Hyvärinen, Information Theory for Systems Engineers. VIII, 205 pages. 4°. 1968. DM 15,20 / US $ 4.20

Vol. 6: H. P. Künzi, O. Müller, E. Nievergelt, Einführungskursus in die dynamische Programmierung. IV, 103 Seiten. 4°. 1968. DM 9,– / US $ 2.50

Vol. 7: W. Popp, Einführung in die Theorie der Lagerhaltung. VI, 173 Seiten. 4°. 1968. DM 14,80 / US $ 4.10

Vol. 8: J. Teghem, J. Loris-Teghem, J. P. Lambotte, Modèles d'Attente M/G/1 et GI/M/1 à Arrivées et Services en Groupes. IV, 53 pages. 4°. 1969. DM 6,– / US $ 1.70

Vol. 9: E. Schultze, Einführung in die mathematischen Grundlagen der Informationstheorie. VI, 116 Seiten. 4°. 1969. DM 10,– / US $ 2.80

Vol. 10: D. Hochstädter, Stochastische Lagerhaltungsmodelle. VI, 269 Seiten. 4°. 1969. DM 18,– / US $ 5.00

Vol. 11/12: Mathematical Systems Theory and Economics. Edited by H. W. Kuhn and G. P. Szegö. VIII, IV, 486 pages. 4°. 1969. DM 34,– / US $ 9.40

Vol. 13: Heuristische Planungsmethoden. Herausgegeben von F. Weinberg und C. A. Zehnder. II, 93 Seiten. 4°. 1969. DM 8,– / US $ 2.20

Vol. 14: Computing Methods in Optimization Problems. Edited by A. V. Balakrishnan. V, 191 pages. 4°. 1969. DM 14,– / US $ 3.90

Vol. 15: Economic Models, Estimation and Risk Programming: Essays in Honor of Gerhard Tintner. Edited by K. A. Fox, G. V. L. Narasimham and J. K. Sengupta. VIII, 461 pages. 4°. 1969. DM 24,– / US $ 6.60

Vol. 16: H. P. Künzi und W. Oettli, Nichtlineare Optimierung: Neuere Verfahren, Bibliographie. IV, 180 Seiten. 4°. 1969. DM 12,– / US $ 3.30

Vol. 17: H. Bauer und K. Neumann, Berechnung optimaler Steuerungen, Maximumprinzip und dynamische Optimierung. VIII, 188 Seiten. 4°. 1969. DM 14,– / US $ 3.90

Vol. 18: M. Wolff, Optimale Instandhaltungspolitiken in einfachen Systemen. V, 143 Seiten. 4°. 1970. DM 12,– / US $ 3.30

Vol. 19: L. Hyvärinen, Mathematical Modeling for Industrial Processes. VI, 122 pages. 4°. 1970. DM 10,– / US $ 2.80

Vol. 20: G. Uebe, Optimale Fahrpläne. IX, 161 Seiten. 4°. 1970. DM 12,– / US $ 3.30

Vol. 21: Th. Liebling, Graphentheorie in Planungs- und Tourenproblemen am Beispiel des städtischen Straßendienstes. IX, 118 Seiten. 4°. 1970. DM 12,– / US $ 3.30

Vol. 22: W. Eichhorn, Theorie der homogenen Produktionsfunktion. VIII, 119 Seiten. 4°. 1970. DM 12,– / US $ 3.30

Vol. 23: A. Ghosal, Some Aspects of Queueing and Storage Systems. IV, 93 pages. 4°. 1970. DM 10,– / US $ 2.80

Bitte wenden/Continued

Vol. 24: Feichtinger, Lernprozesse in stochastischen Automaten.
V, 66 Seiten. 4°. 1970. DM 6,– / $ 1.70

Vol. 25: R. Henn und O. Opitz, Konsum- und Produktionstheorie I.
II, 124 Seiten. 4°. 1970. DM 10,– / $ 2.80

Vol. 26: D. Hochstädter und G. Uebe, Ökonometrische Methoden.
XII, 250 Seiten. 4°. 1970. DM 18,– / $ 5.00

Vol. 27: I. H. Mufti, Computational Methods in Optimal Control Problems.
IV, 45 pages. 4°. 1970. DM 6,– / $ 1.70

Vol. 28: Theoretical Approaches to Non-Numerical Problem Solving. Edited by R. B. Banerji and M. D. Mesarovic. VI, 466 pages. 4°. 1970. DM 24,– / $ 6.60

Vol. 29: S. E. Elmaghraby, Some Network Models in Management Science.
III, 177 pages. 4°. 1970. DM 16,– / $ 4.40

Vol. 30: H. Noltemeier, Sensitivitätsanalyse bei diskreten linearen Optimierungsproblemen.
VI, 102 Seiten. 4°. 1970. DM 10,– / $ 2.90

Vol. 31: M. Kühlmeyer, Die nichtzentrale t-Verteilung.
II, 106 Seiten. 4°. 1970. DM 10,– / $ 2.90

Vol. 32: F. Bartholomes und G. Hotz, Homomorphismen und Reduktionen linearer Sprachen.
XII, 143 Seiten. 4°. 1970. DM 14,– / $ 4.10

Vol. 33: K. Hinderer, Foundations of Non-stationary Dynamic Programming with Discrete Time Parameter.
VI, 160 pages. 4°. 1970. DM 16,– / $ 4.60

PB-36352-SB
5-15